BASIC
COMPUTING
SCIENCE

Logic and Arithmetic in Computing

BASIC
COMPUTING
SCIENCE

Logic and Arithmetic in Computing

John Jaworski B.Sc., M.Sc., M.B.C.S.

BLACKIE

Basic Computing Science
General Editor: W. R. Broderick

System Software	Peter J. Barker
Logic and Arithmetic in Computing	John Jaworski
Data Structures	Brian Bailey
Approaches to Programming	Brian Jackson
Development of Computer-Based Systems	Gail and Paul Sealey
Computer Architecture	Harold Wise

ISBN 0 216 90855 8

First published 1980

Copyright © John Jaworski 1980

All rights reserved. No part of this publication may be reproduced, stored in a retrieval system or transmitted in any form or by any means, electronic, mechanical, photocopying or otherwise without prior permission from the publisher.

Published by Blackie & Son Ltd
Bishopbriggs, Glasgow G64 2NZ
Furnival House, 14–18 High Holborn, London WC1V 6BX

Filmset by Willmer Brothers Limited, Merseyside

Printed in Great Britain by Thomson Litho Limited, East Kilbride

General Editor's Foreword

Computing science is concerned with the processing of information and as such is one of the most fundamental sciences in our society today. This series provides a modern approach to computing science at a level suitable for introductory courses in schools, colleges and universities. It concentrates upon the fundamentals of the subject and provides a strong foundation from which further studies may develop. Both the principles and the practice of the subject are covered. The texts concentrate upon data structures and the ways in which computers manipulate data. Processors and peripherals are seen as tools used for collecting and processing data. Thus, there is reference to computers in terms of their logic, architecture and characteristics but there is no undue concentration upon details of how particular pieces of equipment work. A knowledge of the fundamentals of programming is essential to computing science and the emphasis in this series is upon the concepts rather than on the detailed explanation of modes of operation. Programming practice is considered an essential element of data processing and where necessary and desirable, the texts are illustrated with examples in common computing languages.

Students working at this level find it necessary to relate the theory to practice, and opportunities are taken to illustrate the algorithms and practices derived from the theoretical study. The problems of high volume data processing operations and other applications are discussed.

Whilst the approach is academic it is related to the practice of computing wherever possible. It is aimed (in the United Kingdom) at GCE Advanced Level and at First Year Higher Education Service Courses, but the breadth and depth of coverage is such as to make the series immediately useful to those with a serious interest in computing science and its practice.

Logic and Arithmetic in Computing deals with both the theory and practice of logic and arithmetic in a computer system. Its scope is unusually wide and ranges from a study of basic logic gates, through format and notation for numbers, to the design and implementation of mathematical subroutines for such activities as random number generation and evaluation of trigonometric functions. The style and approach that John Jaworski has adopted has avoided the normal arid treatment of this topic and has produced a readable and enjoyable text which will lay a strong foundation upon which the reader can build if necessary.

W. R. Broderick

Author's Foreword

This book can be deceptive. The major trap lies in the fact that pages of print look about the same, whether they are carrying densely-packed, tightly-reasoned argument, or light, inoffensive commentary. For example, in Chapter 1, Section 1.1 (page 3), there is a table setting out (a part of) the 256 possible 8-bit patterns. It is not necessary to do any more than to glance at this table and to observe its regularity and basic structure. But in Section 1.2, on page 5, we observe the line:

1 0000000 or −1 (!)

This is a line dense with information—the exclamation mark indicates that we might be moderately surprised by this statement, even that we should check the calculation. There is no defence against this variety. This is the only line in the book with a warning '!' and the reader is encouraged towards discrimination and caution.

The chapters are not of equal density either. By-and-large, the first four chapters are a fairly light narrative romp through the topics described—this should be enjoyable, and why not, indeed? But Chapter 5, and to a lesser extent Chapters 6 and 7, are more overtly mathematical than their predecessors; they are, quite simply, harder going. The material therein needs a little more work to appreciate it. Where solutions to the exercises can be presented reasonably concisely, they are to be found in the appendix.

It is customary to acknowledge those among one's colleagues who have provided help and encouragement. This book is a collaborative venture; without the good offices of Bill Broderick, it could never have happened. Meurig Williams, Judith Anderson and John Turnbull may no longer recognize the work they criticized, but I thank them warmly nevertheless. It is not clear exactly what Bill Tagg contributed to the manuscript, but in employing me in a capacity where fruitful and engaging discussion of these topics was possible, he lays a claim to being the godfather of the book. For the material in Chapter 5, I must thank P. H., my favourite computer pirate, who supplied me with a number of source-code listings of compilers. (Some computer managers do not know what software they possess!) I also wish to thank the University of Cambridge Local Examinations Syndicate, the Associated Examining Board, the Oxford Delegacy of Local Examinations and the University of London, University Entrance and School Examinations Council for permission to reproduce questions from past examinations.

Finally, it is customary to dedicate a book. I should like to dedicate Chapters 1 to 4 to my girl-friend, and Chapters 5 to 7 to my wife. By dint of some non-numeric processing that we effected between chapters she is one and the same person. There is however a warmer and more personal dedication hidden in this book. To find it will involve the reader in some non-numeric processing of her own, but if she should detect some unexplained discrepancy within these pages, she may be confident that she has found it.

John Jaworski

Contents

1 The Representation of Numbers

Think of a number. You might have thought of 7. But you could have thought of −7, $3\frac{1}{4}$, or 0 or 1 000 001. The computer cannot be as selective as this. It must be able to process and store numbers like any of those above. That is, it must be capable of dealing with:

1. positive and negative whole numbers (integers);
2. fractions, and numbers with fractional parts;
3. very large (and very small) numbers;
4. zero.

This must all be done within a consistent and coherent framework. The user of the computer will expect only a few, obviously sensible, restrictions on what numbers he is allowed to use, and how he writes them.

Consistency demands that we use the same basic tools to represent each of the classes of number listed above. That is, we shall use computer *words* to maximum efficiency, try as far as possible to associate one computer word with one number, try to process one computer word at a time, and try to use those instructions in the computer's machine code that are easy and economical to provide.

Throughout the chapter, we shall assume that our computer has its main storage arranged in *words*, each of 8 *bits*. The extension to other arrangements and other word sizes is a simple matter. Another book in this series—*Data Structures* by Brian Bailey—discusses how computer storage is used to encode information.

1.1 Representation of integers in the computer

An integer is simply a whole number with no fractional part. It may be positive or negative, or even zero (which is neither positive nor negative). Outside the computer we write integers as

+7 −87 43 1,001 0 and so on.

We normally assume that an integer without a sign (e.g. 43) is positive, unless it is 0. In what follows we shall adopt the convention that all integers will be written with a sign (+43, −87, etc.) even where it would be usual to omit it, and further, that integers will not be punctuated.

We have 8 bits at our disposal; this allows us to represent a maximum of $2^8 = 256$ different numbers. Written out systematically, these bit-patterns would be:

00000000
00000001
00000010
00000011
00000100
.. .. ..
11111110
11111111

It is tempting to associate these with the integers 0, +1, . . ., +255 in the obvious way suggested by binary arithmetic.

Life gets a little more complicated when we think of how we should represent the sign (+ or −) of an integer. There are three possible ways.

Sign-and-magnitude representation

Without straying outside the 8 bits allowed, it seems reasonably clear that information about the + or − sign will use up one of our bits, say the leftmost one. It seems natural to do what we do in ordinary print, which is to write down the sign, and then to follow it with the 7-bit representation of the number itself, that is, to consider the word divided as in Fig. 1. It is customary to use a 1-bit to represent a − sign, and a 0-bit a + sign.

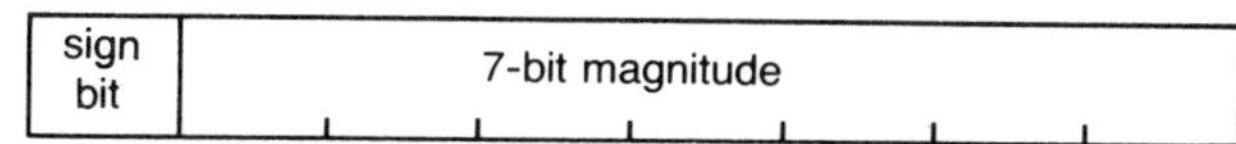

Fig. 1 Sign-and-magnitude representation

There are only 7 bits left to represent the number, and these will allow 2^7 or 128 distinct representations, from 0 to 127. Our bit patterns and their corresponding integers (arranged more naturally now) will be:

```
1 1111111     -127
1 1111110     -126
1 1111101     -125
. . .. ..      . .
0 1111101     +125
0 1111110     +126
0 1111111     +127
```

Almost at once we spot that something has gone slightly wrong. From −127 to +127 (inclusive) there are just 255 different numbers (there are 127 negative numbers, 127 positive numbers, and 0) Where did we lose one? If we look at the middle portion of our list, we can see:

```
. . .. ..      . .
1 0000010     -2
1 0000001     -1
1 0000000     -0 ?
0 0000000     +0 ?
0 0000001     +1
. . .. ..      . .
```

Our method of representation forces us to attach a sign to every integer, and so we are faced with +0 and −0. Just by itself, this isn't a great disadvantage. Admittedly, we are not being as efficient in the use of our storage as we might be, but losing one number in 256 is a loss of only 0.4%. However, there are other more substantial disadvantages.

Our output routines must now take care that −0 does not appear in the print-out (+0 will be taken care of, along with all positive integers). Every time that a negative integer is encountered, an additional check must be made to ensure that the number to be printed is not 1 0000000, which would appear as −0. The − sign must be suppressed.

Also, comparison with zero (in an instruction like JUMP IF ZERO) becomes more difficult, because there are two valid representations of 0. Indeed, JUMP IF NEGATIVE becomes slightly harder, as we cannot simply say that a number is negative if its binary pattern begins with a 1. However, JUMP IF POSITIVE has a similar and unavoidable problem.

The most serious, but least obvious disadvantage arises when we wish to actually carry out calculations with numbers. Consider the three addition sums set out (in denary) below.

```
+347      -347      +347
+243+     -243+     -243+
-----     -----     -----
+590      -590      +104
```

Because numerically the result of the first two problems is the same (and the sign an afterthought) it is likely that the same algorithm is used to perform the numerical part of the calculation. But this cannot be predicted in advance when the program is being compiled, because the value of the operands, and in particular their signs, cannot be known until execution time. The third addition, although using similar operands, is rather different in character, and is essentially a subtraction. Two major implications arise from this. We must test the signs of the operands at execute time, and a separate algorithm is required for subtraction (it cannot be seen as negation, followed by addition). More ingenious methods are required!

2's-complement representation

Let us consider one particular 8-bit pattern:

11111111

Whatever it represents, it is clear that treating it as a binary number (it looks like +255, but we shall resist calling it that for the moment) and adding 1 to it gives us the 9-bit pattern

100000000

However, we are restricted to 8 bits, and we might like to consider the left-hand 9th bit 'falling off the end' leaving us with the 8-bit pattern

00000000

Whatever system of representation we adopt, it seems clear that this pattern is an excellent candidate for 0. In which case, the original pattern, 11111111, is equally an excellent candidate for −1. It is, after all, the number which 1 must be added to, to give 0.

We can clearly extend this idea (just a little way for the moment).

Looking at the numerically small positive and negative integers and arguing in a similar way, we get:

.. .. ..	..
11111101	−3
11111110	−2
11111111	−1
00000000	0
00000001	+1
00000010	+2
00000011	+3
.. .. ..	..

There is an immediate attraction in this scheme. There is just one 0 representation, and it is the natural one, 00000000. Further, it looks as if positive numbers (and 0) all have a leading 0-bit, while negative numbers have a leading 1-bit. In some sense, the leftmost bit is functioning as a sign-bit, even though it is also involved in carrying numerical information, as we saw when we added 1 to −1. It seems unlikely that there will be any double representation (no +0 and −0) so we can expect a full 256 different numbers from our 8 bits.

There is just one problem, but it is easily resolved. We can see what the problem is, by extending the table slightly. If we count on, in the positive direction, we get to:

.. .. ..	..
01111101	+125
01111110	+126
01111111	+127
10000000	?

What does 10000000 stand for? As we got it by adding 1 to 127, it seems to be a plausible +128, except that we could also count backwards and get:

.. .. ..	..
10000001	−127
10000000	−128 ?

The problem is that large negative numbers have the same representation as small positive ones, and vice versa. If we wish to preserve that convenient leading sign-bit, it is clear that the 128 patterns beginning with 1 are to be considered negative, the rest 0 or positive. The full table is then:

10000000	−128
10000001	−127
10000010	−126
10000011	−125
.. .. ..	..
11111101	−3
11111110	−2
11111111	−1
00000000	0
00000001	+1
00000010	+2
.. .. ..	..
01111101	+125
01111110	+126
01111111	+127

Some further observations are in order. First, 10000000 (−128), the most negative number, does not have a corresponding positive number (+128). This could have a minor effect, in that the computation of an expression like

$$(-1)-(-128)$$

could result in an arithmetic error, if the compiler attempts to turn it into

$$(-1)+(+128)$$

before evaluating it. That this problem arises at all is due to another convenient property of 2's-complement representation, namely that there is now no need to have a separate algorithm for subtraction. Consider the subtraction

$$(+5)-(+3)$$

If the compiler turns this into

$$(+5)+(-3)$$

by simply negating the (+3), and then treats the bit-patterns as unsigned binary numbers, we get

+5	00000101
−3	11111101
+2	(1)00000010

where the bracketed (1) indicates the extra 9th bit 'falling-off'. This process actually has more pitfalls than the above suggests, but we shall return to it in Chapter 3.

Another observation is that all even numbers now end in 0 and all odd numbers in 1. This is not particularly important but is another instance of the cunning involved in the 2's-complement notation, where we can choose to ignore the sign information at times convenient to us.

Two's-complement arithmetic can be initially rather puzzling: it is not at once apparent how to read off what a negative number is, but this becomes easier with practice.

1's-complement representation

There is one final method of representing signed integers that must be included for completeness, although it has few of the advantages of 2's-complement.

Assume that we use the 128 patterns which begin with a 0 to stand for the integers 0 to +127 (as in both previous methods). We could efficiently convert to the negative numbers by changing each 0 to a 1, and each 1 to a 0. The leading bit will be changed to 1, indicating a negative integer. The full list is given below, together with the corresponding 2's-complement list.

decimal	*1's-complement*	*2's-complement*
−128		10000000
−127	10000000	10000001
−126	10000001	10000010
−125	10000010	10000011
..	.. .. ..	.. .. ..
−3	11111100	11111101
−2	11111101	11111110
−1	11111110	11111111
−0	11111111	
+0	00000000	00000000
+1	00000001	00000001
+2	00000010	00000010
..	.. .. ..	.. .. ..
+125	01111101	01111101
+126	01111110	01111110
+127	01111111	01111111

Notice that we have reintroduced +0 and −0, and lost the odd/even property. However, compare the 1's-complement list with the 2's-complement list. For negative integers, the 2's-complement is obtained by adding 1 to the 1's-complement. This gives us an easy method of negating a number in 2's-complement form (see Fig. 2).

	Example 1 −(+5)	Example 2 −(+127)	Example 3 −(−7)
(1) take the number	00000101	01111111	11111001
(2) form the 1's-complement	11111010	10000000	00000110
(3) add 1	11111011 (−5)	10000001 (−127)	00000111 (+7)

Fig. 2

As Example 3 shows, the method even works with a negative integer. Forming the 1's-complement is a relatively easy step which is a contribution to the success of 2's-complement notation.

Of course, in using the full information-carrying capacity of an 8-bit computer word to represent integers, we have used up all the different bit patterns that can be formed. Whatever methods we adopt for the representation of fractions, large numbers and so on, the patterns we obtain will look like some integer representation. This is unavoidable. We must ensure that whenever we process the computer representation of any quantity, we understand what it represents. We shall get very strange results from applying integer algorithms to fractional representations.

1.2 Representation of fixed-point fractions in the computer

A *fraction* is any number between 0 and 1. If written out in denary, it would be 0.5, 0.878326, 0.111, 0.001 or something similar. It is initially attractive, and quite productive, to think of storing fractions in a way similar to integers, by using the place-values of the binary system. Thus, in binary:

0.1 would represent 2^{-1} or 0.5
0.111 represents $2^{-1}+2^{-2}+2^{-3}$ or 0.875
0.0001 represents 2^{-4} or 0.0625
and so on.

We do not need to store the point explicitly, although we shall use a sign-bit. The three fractions above would appear as

0 1000000 (0.5)
0 1110000 (0.875)
0 0001000 (0.0625)

and, using 2's-complement representation, their negations would be

1 1000000 (−0.5)
1 0010000 (−0.875)
1 1111000 (−0.0625)

This system of representing fractions is called the *fixed-point* method, because it uses a point fixed in position, at the same point in each word. We may think of it at the left, although it obviously comes between the sign-bit and the most significant bit of the numerical information.

Using fixed-point fractional representations, the largest and smallest positive fractions representable in 8 bits are

0 1111111 or 0.9921875
and 0 0000001 or 0.0078125

Because of the imbalance between positive and negative in the 2's-complement notation, the most negative number representable appears as

1 0000000 or −1 (!)

and the closest to zero:

1 1111111 or −0.0078125

We can also see 0 as a fraction, if we wish. Depending on our system (1's-complement, 2's-complement, etc.) we may or may not be faced with +0 and −0. The details are left to the reader.

There is a sense in which the integer representations discussed in Section 1.1 are also fixed-point representations, in that an understood point is to be found at the right-hand end of the word. In fact, the term can be applied to any notation where the point takes up a fixed position.

There are two important considerations to bear in mind when dealing with fixed-point fractions.

Precision

While it is possible to represent all the integers in a given range exactly with a sufficient number of bits, this is not true for the fractions. This is not unexpected, as we cannot even write these down. Between 0.0078125 and 0, for example, there are a substantial (indeed infinite) number of fractions that do not have a representation. When dealing with fractions, we shall have to approximate some by using a suitably close fraction that does have a representation. This problem is theoretically unavoidable. The more bits we have, the better our approximations can be, but the need for approximation cannot be removed. All fractional representations run the risk of being imprecise.

Multiplication

When any 8-bit quantity (integer or fractional) is multiplied by another 8-bit quantity, the resulting product will always be 16 bits long. (This is always true in any base. The product of 2, n-digit quantities is $2n$ digits long.) With integers, the product cannot be stored in an 8-bit word, and it is not clear how an approximate answer may be derived (in fact, this is impossible unless floating-point methods, to be discussed later, are employed). Of course, coincidentally, the first 8 bits of the product may all be 0's (1's for a complementary representation), but this cannot be guaranteed, and the first 8 bits cannot in general be ignored.

With fractions however, although a 16-bit product is produced, it is possible to ignore the lower-order 8 bits (which represent greater precision) and use the higher-order 8 bits as an approximate product. This is very convenient, particularly in dealing with floating-point representations (see Section 1.3).

1.3 Extending the range of our representations

Having 8 bits per word is a good discipline. We cannot have a number larger than +127, nor a positive fraction smaller than 0.0078125. But computers habitually perform calculations that are both literally and figuratively astronomical. How are we to use the limited tools at our disposal to extend our range of numbers?

It is unrealistic to try to solve this problem with just 8 bits. We need some more, and a convenient way to get them is to annex a second word, giving 16 bits in all. One way of using this extra capacity is to rework the previous sections using 16-bit words. This will still not be enough, although it is a solution that may be adopted. Instead, we shall look at another method of representing numbers that extends our range at the expense of precision.

Floating-point representation

Let us take a lesson from the (imperfectly implemented) metric system of measurement. This has a hierarchy of units, each a ten-fold multiple of its predecessor.

millimetres	mm
centimetres	cm
decimetres	dm
metres	m
decametres	dam
hectometres	hm
kilometres	km

For a variety of eminently practical reasons, many of these terms are not in everyday use, even among scientists. However, it is possible to represent any measurement within the range of this hierarchy, without using a decimal point. Thus, on halving a measurement of 1 km, it is unnecessary to use 0.5 km. Instead we could write 5 hm. Halving again, 2.5 hm is unnecessary, as this could be written as 25 dam. Repeated halving, with appropriate choice of units, gives

125 m 625 dm 3125 cm and so on.

The principle here is that it is possible to represent measurements in a consistent way as a fixed-point quantity (here an integer), together with the appropriate unit.

This is the way we shall represent a wider range of numbers in our computer, now augmented to 16 bits/number. We shall use the second word of each pair to carry information about the unit, and the first word to hold a suitable fixed-point representation. We have a choice of storing either integers or fixed-point fractions in the first word of the pair, and initially there is little to choose between them. The second word of the pair will be adjusted to make the quantity 'come out right' whichever we choose.

In fact we come down in favour of fixed-point fractions. The reason is the resulting simplicity in processing. Recall that we observed that *fraction* × *fraction* = *fraction*. This means that the product of two fixed-point fractions each of 8 bits gives a fraction that is initially longer than 8 bits, but which can be simply truncated to 8 bits. This involves a loss of precision, but leaves us with an approximate answer. To achieve the same with integers, it would be necessary to retain more than 8 bits and to shift the bits to ensure that we did not lose any of the significant bits. The point at the left-hand end is a valuable watchdog to make sure that we need not worry about this.

We now only need to agree on how to represent the information in the second word of the pair, and we can start looking at some *floating-point representations*. They are so named, because we have abandoned attempts to explicitly represent the point, which now 'floats' implicitly in a position given by the unit. It is tempting to refer to such quantities as floating-point numbers but this is inaccurate, as it is the representation that has the floating-point, not the number itself.

We take our clue for representing the units from the metric system again. Each prefix (kilo-, hecto-, etc.) stands for a power of 10.

milli-	10^{-3}
centi-	10^{-2}
deci-	10^{-1}
	10^{0}
deca-	10^{1}
hecto-	10^{2}
kilo-	10^{3}

The 10 is the base: it need not be stored, as it is always the same. In a binary computer, the base would be 2, and similarly it would be understood rather than explicitly stored. We need only consider how to store the exponent, and as this will always be integral, we arrive at the scheme shown in Fig. 3.

Every number has many representations in a floating-point system, and we need a convention for selecting one of these as standard, if only to simplify comparison in a statement such as JUMP IF EQUAL.

(In practice it is very unwise to use such a comparison test on two floating-point representations, or indeed two fractions, especially where one or both have been obtained as the result of calculation. Rounding errors, and the inherent loss of precision mean that only rarely are two 'mathematically equal' numbers equal in the computational sense. It is better to ask if they are 'close'—that is, if their difference is less than some pre-defined tolerance reflecting the probable error.)

We select for our standard representation the one that also gives us the maximum amount of precision, effectively the maximum number of significant figures. That is, we choose the representation that places a

Fig. 3 Floating-point representation

Let us look at a few examples. As we shall refine this method in two distinct ways in the next section, we shall not spend too long here. We shall assume the 2's-complement method in both words.

0 1000000	0 1000000	is 0.5×2^{64} (or 2^{63})
1 1000000	0 1000000	is -0.5×2^{64} (or -2^{63})
0 1100000	1 1000000	is 0.75×2^{-64} (3×2^{-66})

this clearly demonstrates the vastly increased range. However, we can do rather better in a couple of respects.

Normalized floating-point representation

We can improve on the floating-point method in two ways. There is a basic problem that can be seen in looking at the following patterns.

0 0000001	0 0000111	($= 2^{-7} \times 2^{7}$; that is, 1)
0 0000010	0 0000110	($= 2^{-6} \times 2^{6}$; 1 again)
0 0000100	0 0000101	($= 2^{-5} \times 2^{5}$; 1 again)
0 0001000	0 0000100	($= 2^{-4} \times 2^{4}$; 1 again)

1-bit in the first position after the sign-bit. Such a representation is called a *normalized* floating-point representation. For example, consider

0 1110110	0 0001011	(normalized)
0 0011101	0 0001101	(unnormalized)

The two underlined bits in the normalized representation are information of significance that has been lost in the unnormalized case.

The algorithms discussed in Chapter 4 usually produce unnormalized intermediate results. The last step in a floating-point arithmetic operation is usually to normalize again, implying that such operations are carried out in registers capable of holding a greater number of bits than can eventually be stored.

Zero cannot be normalized, and the most obvious convention is generally adopted:

0 0000000 0 0000000

We have already noted that the second word of each pair (the *exponent*) is always an integer. As the value of this integer is only required by the floating-point hardware or software and is never directly accessed by the programmer, it is usually stored in a form convenient for the hardware. In our example, using 2's-complement storage, the range is apparently −128 to +127. It is conventional to add 128 to this (effectively treating the exponent as an unsigned integer).

We treat the bit-patterns from 00000000 through to 11111111 as positive integers from 0 through to +255, and the hardware (say) adjusts by subtracting 128 automatically. Many algorithms which compare or adjust the exponents in two words can work efficiently with the unsigned version (the excess cancelling out). Such a storage convention is known as *excess-128 storage*.

It is not at once clear how normalization affects negative numbers (that is, those with a negative mantissa). Shifting these to the right, when they are apparently normalized will produce a representation that is still apparently normalized. Thus

1 1110101 0 0000101
1 1111010 0 0000110
and 1 1111101 0 0000111

all represent the same number (precision aside). In fact, the description of normalization given above is only correct for positive quantities, but is easily generalized. Normalization places a significant bit to the right of the sign-bit. With negative numbers, the significant bits are 0-bits, not 1-bits. So a normalized negative number is one where the first bit after the sign-bit is 0. A more formal definition of normalization is thus: choosing the representation where the first bit after the sign-bit is different from the sign-bit (at least for the complement notations).

Consider the 1's-complement negative number 1 1111010. Because of the way 1's-complement is formed, we 'read significance' into the 0-bits, mentally evaluating the positive number 00000101 ($= 2^2 + 2^0$, or 5) and attaching a minus sign to it. For 2's-complement the situation is a little more tricky, as we attach significance in the same way to zeros, up to the rightmost 1-bit, and attach significance to that. That is, we read significance into the underlined bits in 11111010 to give −6. Sign-and-magnitude representation, where the magnitude is always positive, avoids this complication entirely, and is normalized as for a positive number.

1.4 The range of normalized floating point representations

To see how much we have gained in using this convention, we shall evaluate some large and small representations.

The largest floating-point representation

This is given by the pair of words

0 1111111 11111111

The *mantissa* is positive, and is 0.9921875. It is more useful to note that the mantissa is the binary equivalent of the denary 0.9999999, or approximately 1.

The *exponent* is the unsigned integer 255, which in the excess-128 convention represents the exponent +127. This number is thus

$$0.9921875 \times 2^{127}$$

which is of the order of 1.7×10^{38}. Denary exponent notation does not express the magnitude of this number very well. It is about

170 000 000 000 000 000 000 000 000 000 000 000 000

or rather more than the number of millimetres in

50 000 000 000 000 000 000 000

trips to the sun and back! This is a considerable improvement over +127.

The smallest positive floating-point representation

This is given by the pair

0 1000000 00000000

That is, the smallest possible normalized mantissa and the smallest exponent. Remembering that the exponent is in excess-128 notation, we get

$$0.5 \times 2^{-128} \qquad \text{which is } 2^{-129}$$

In denary, this is of the order of 1.5×10^{-39}

The numerically largest negative floating-point representation

To avoid confusion we shall use 2's-complement throughout. We are seeking a large negative mantissa and a large positive exponent. This is given by the pair.

1 0000000   11111111

or, in denary

-1×2^{127} approximately -1.7×10^{38}

The numerically smallest negative floating-point representation

Remembering that we require a normalized mantissa, we get the pair

1 0111111   00000000

which is, in denary

$-0.5078125 \times 2^{-128}$ approximately -2.9×10^{-39}

Recall that we attach significance, in the mantissa, to the leading (only) 0-bit and the rightmost 1-bit. The mantissa is thus $2^{-1} + 2^{-7}$.

1.5 Precision in floating-point representations

The increase in range that floating-point representations allow, is at the expense of precision.

In considering precision, we can profitably ask how far apart the adjacent representations are. If the difference is small, precision is greater than if it is large. However this difference varies, depending on the size of the number being represented. For example, the largest representable number is given by the pair

0 1111111   11111111

representing approximately 2^{127}. The next nearest number to this that can be represented exactly involves changing the least significant bit of the mantissa from a 1 to a 0, giving

0 1111110   11111111

The actual magnitude of the change is influenced by the size of the exponent. It is a change of 2^{-7} in the mantissa, but is multiplied by the exponent, 2^{127}. It is thus a net change of 2^{120}. At the upper end of our range, adjacent quantities are at a distance 2^{120} apart. Precision is very low here.

At the numerically smaller end of the scale, representations are packed very much more closely. The two smallest positive representations are

0 1000001   00000000
0 1000000   00000000

which differ by $2^{-7} \times 2^{-128}$. Precision is very high indeed, as we can 'resolve' numbers as close as 2^{-135} apart.

Precision can obviously be improved somewhat by allocating more bits to the mantissa but, however much storage is given over to the mantissa, precision will still be greater where the exponent is small.

1.6 Packed representations

Our use of two 8-bit words in floating-point representation is convenient in that it demonstrates the principles. However, the practical details are usually modified slightly in computers with a longer word length, or those organized on a *byte* basis (effectively a variable-length word basis).

For example, if our computer had 16-bit words, it would have been necessary to unpack the mantissa and exponent (by means of masking and shifting operations as described in Chapters 2 and 3). There is no especial virtue in using 8 bits for both fields, and we could well choose a 12-bit mantissa and 4-bit exponent, say, or any other division that proves to be a useful compromise between range and precision. It is customary to give the exponent less space than the mantissa.

Summary

In this chapter we have seen how we can use bit patterns within the computer to represent all common classes of numbers. Integers (whole numbers, without fractional parts) may be represented exactly over a limited range in any of the following ways:

1. unsigned (with 8 bits, the range is from 0 to +255)
2. sign-and-magnitude (−127 to +127)
3. 2's-complement (−128 to +127)
4. 1's-complement (−127 to +127)

The advantages and disadvantages of these methods are discussed in Section 1.1.

Fractions can be represented in four similar ways (unsigned fractions are not discussed in this chapter). Only a limited degree of precision is possible.

The integer and fraction representations are fixed-point representations, in that the (binary) point is implicit, in a fixed position.

Floating-point representation may be used for larger integers, for mixed numbers and for numerically small fractions. This usually stores a fixed-point fraction and an exponent (often stored in a coded form known as excess-n storage). In order to preserve the maximum possible amount of precision, floating-point numbers are usually normalized. That is, a representation is chosen that has a significant figure after the implied point.

EXERCISES

Chapter material

These exercises are intended as fairly short questions to test your understanding of the material in the chapter. Answers are given at the end of the book.

1. *Using 8 bits/word, and fixed-point integer storage conventions, how would each of the following be represented in*
   - *a. sign-and-magnitude,*
   - *b. 2's-complement, and*
   - *c. 1's-complement form?*

   (i) +83 (ii) −83 (iii) −1 (iv) +31 (v) −96
2. *Using the same conventions as in Question 1, what integer would each of the following represent in each of a, b, and c?*
   (i) 11111111 (ii) 01111111 (iii) 10000000
   (iv) 10101010 (v) 11001101
3. *Using 12 bits/word, and the conventions of the above questions, what are the largest and smallest (most negative) integers representable, using each of a, b and c?*
4. *How large a word-size is necessary to represent 1000 different numbers?*
5. *What does each of the bit patterns in Question 2 represent if it is to be interpreted as a fixed-point fraction with a 'binary point' following the sign-bit? Give an answer for each of a, b and c.*
6. *Using 2's-complement form, and 8 bits/word, how would each of the following be represented?*
   (i) 0.75 (ii) 0.625 (iii) 0.03125
   (iv) −0.625 (v) −0.5
7. *Which is the best 8-bit pattern to represent each of the following in 2's-complement form?*
   (i) 0.1 (ii) 0.3 (iii) 1/3 (iv) −0.1 (v) −1/5
8. *Using 16 bits, divided as a 12-bit mantissa and a 4-bit exponent (with excess-8 storage of the exponent), what is:*
   - *a. the largest positive representable number, in floating-point form;*
   - *b. the smallest positive representable number;*
   - *c. the largest negative representable number;*
   - *d. the smallest negative representable number?*

   Use 2's-complement form for the mantissae.
9. *Using a pocket calculator or a computer, give your answers to Question 8 as powers of 10.*
10. *Using the floating-point convention of Fig. 3, what do each of these patterns represent?*
    - *a. 0 1101101 10101010*
    - *b. 1 0001101 11111111*
    - *c. 1 0001111 00011101*

Public examination questions

These exercises are taken from public examinations, and are intended to give practice in recognizing how questions on the material in this chapter are posed in practice. For guidance, the marks awarded for a fully correct answer, and the length of the examination paper are given; thus for a question marked 6%, 3 hours, it might be reasonable to spend 6% of 3 hours (10–11 minutes) in answering. No solutions are given to these questions.

11. *Write down the binary digit patterns that would result in a 12-bit word computer from the storage of +31 and −31 in each of the following cases:*

a. *two's complement storage of negative integers,*
b. *one's complement storage,*
c. *explicit storage of the sign, with the convention that 1 indicates a negative number.*

(London 1975) (6%, 3 hours)

12. a. *What is the numerically largest negative integer that can be represented by 8 bits, using 2's-complement fixed-point storage? Give your answer in binary and in denary.*
b. *The floating-point representations of numbers are usually normalized. Why?*

(London 1977) (4%, 3 hours)

13. a. *Define the two's complement of an integer X in an n bit word and give a simple practical method of writing down this complement for any given number.*
A register has a word length of a sign bit followed by nine bits. Display the greatest and least positive and the greatest and least negative integers and fractions it can hold (exclude zero) using two's complements and state the decimal value of each number displayed.
Using the above word length and two's complements evaluate (i) −21 +2; (ii) 256 +256; (iii) (−1/4) +(−1/2) and comment on your results where necessary.
b. *Express each of the following decimal numbers in the decimal floating point form $A \times 10^n$ where $0.1 < A < 1.0$ and n is a positive or negative integer.*

4796.0 3.999 0.0913

Without rounding at any stage add these floating point numbers in the order given. Show that a more accurate answer can be obtained and comment on this.

(A.E.B., 1974) (17%, 3 hours)

14. *The word length of a certain computer is sixteen bits. State the range and accuracy of numbers which can be held in a store location as (a) integers, (b) fractions and (c) normalized floating point numbers in the format shown in the diagram at the top of the next column, using two's complement form for negative numbers in both the mantissa and exponent.*

Sign bit	*10 bit mantissa*	*Sign bit*	*4 bit exponent*

Show how the decimal numbers 372.25 and −186.125 would be displayed in the above register in normalized floating point form. Add these two numbers in this form showing each step and leaving the answer in normalized form. Ignore round-off at all stages. Explain an alternative method for representing the exponent which avoids the use of negative numbers.

(A.E.B., 1975) (17%, 3 hours)

15. *A certain number is stored in floating point form in 16 bits as follows:*

0 1 0 1 0 0 0 0 0 0 0	*0 0 0 1 0*
A	*E*

A is an eleven bit fraction in two's complement form, with the binary point after the leftmost bit.
E is a five bit binary exponent in two's complement form.
The value of the number is $A \times 2^E$.

a. *What is the value of the number shown above?*
b. *Explain what is meant by the normalizing of numbers stored in floating point form, and why this is done.*
c. *What are the largest and smallest normalized positive values that can be represented? (Express your answer in powers of two).*
d. *How many binary digits of accuracy can be represented, and approximately how many decimal digits is this equivalent to?*

(Cambridge Specimen Paper) (17%, 3 hours)

16. *How many different numerical quantities may, at most, be represented by using 8 bits?*
State the range of integers that may be represented by using 8 bits, and the 2's complement convention.
Show how this range may be extended, using a floating-point convention.

(London, 1979) (6%, 3 hours)

Projects

These exercises are suggestions for short projects exploring your own computing environment.

17. *By assigning numerical quantities to FORTRAN variables using F and I formats, and printing out the binary patterns using 0 format, discover the particular conventions used in your system. Check this using a manual, and by inputting octal data and printing it using F and I format. A driving program to allow you to input bit patterns might be useful.*
18. *Can you discover at which point your BASIC system switches over to E format for printing out numbers? Is there any way in your system to discover or inspect the binary patterns?*

2 Designing the Computer's Logic

Understanding how a computer is built is of no concern to a computer scientist! His only interest is the effect that the components of the computer have on the data.

Data flows through a computer as strings of 1-bits and 0-bits. These might be represented electronically (by voltages, say), or even by temperatures. We can simply think of the bits flowing by themselves. The whole system is synchronized by a clock, which drives the system in such a way that these bits arrive when they are needed, rather than haphazardly.

Each component of a computer system has a number of *inputs*, and our interest begins at the time when a bit arrives at each of the inputs simultaneously. Thus, each input is either set at 0, or set at 1. After a time delay, due to the actual working of the component, the *outputs* of the component will also be set at 0 or set at 1, depending on the effect that the component has had. A component may have as few as one, or as many as several thousand inputs and a similar number of outputs. However, it is usually the case that components with large numbers of inputs and outputs can be considered as being assembled from a number of smaller components, each with rather fewer inputs and outputs. Therefore, it is worth studying fairly exhaustively those components with a few inputs and a few outputs that experience tells us are useful as the basic building-blocks of the computer. That is what this chapter does. Our approach will be to treat these components as 'black boxes'—ignoring entirely any possible circuitry or electronics between the inputs and the outputs. We can characterize each component solely in terms of what it does to the inputs (i.e. what outputs it produces for each set of inputs). This will, at the very least, save us from being out-distanced by rapidly changing technology that would leave this text old-fashioned before the ink was dry on the page!

The simplest components—black boxes that is, are often called *gates*, because their action can be seen in terms of permitting a signal to pass (setting a 1-bit at an output) or preventing one (setting a 0). The term is used rather loosely, and often just means any logic circuit. We shall try to use it for those components with a single output, where the term fits most naturally.

2.1 Gate circuits

The NOT gate

The simplest component is one with a single input and a single output. If the output is to depend on the input, it is reasonably clear that the output will either be the same as the input, or different. If the output is the same as the input, the component can hardly be said to be doing much (beyond introducing a timing delay), so the only practicable component of this simplicity is the NOT gate, whose action is defined by the table in Fig. 4.

INPUT	OUTPUT
1	0
0	1

Fig. 4

It gets its name because the output is related to the input by the mathematically logical operation of NOT-x.

How many 1-input, 1-output gates are there (useful or not)? We can answer this by asking how many different tables like the one above could be drawn up. There are four in all, the one above and the three in Fig. 5. Tables (1) and (3) of Fig. 5 ignore the input altogether, maintaining a 1 output or a 0 output respectively, while the second passes the input signal unchanged. This second gate does introduce the time delay mentioned above, so it is not entirely without use, but timing considerations are the province of the computer engineer, so we shall not develop the study of components like these any further.

INPUT	OUTPUT
1	1
0	1

1.

INPUT	OUTPUT
1	1
0	0

2.

INPUT	OUTPUT
1	0
0	0

3.

Fig. 5

We represent the NOT gate by the symbol in Fig. 6. Many people use special symbols for gates such as these, each gate having its own shape. Unfortunately, many different conventions are in use, and there is far from universal agreement as to which is best. We shall adopt a neutral posture, and simply represent any component by a square box, with an appropriate label.

Fig. 6

Notice that our approach does not allow for the possibility that the gate reacts to the input in some way that is determined by previous inputs—by the 'history' of the gate. We shall look at some components that adopt another method later in this chapter.

The AND gate

The next simplest gate has two inputs (and a single output); in theory, there are 16 possible gates of this type, only 6 of which are of practical use. One of these is the AND gate, represented by the symbol and defined by the table in Fig. 7.

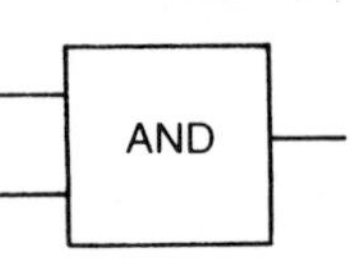

INPUTS	OUTPUT
0 0	0
0 1	0
1 0	0
1 1	1

Fig. 7

In common with the other 5 gates of this type, both inputs are treated with equal weight. If one input is 1 and the other 0, the output is the same (0) regardless of which input is set at 1 and which at 0.

The NAND gate

Fig. 8 defines another 2-input gate, whose output is simply that which would be obtained by applying the output of an AND gate as the input of a NOT gate. Hence NOT-AND, or NAND for short.

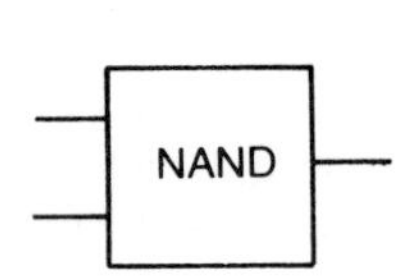

INPUTS	OUTPUT
0 0	1
0 1	1
1 0	1
1 1	0

Fig. 8

There is a certain academic fascination in the NAND gate, for we can show that it is possible to build all gates from a suitable combination of NAND gates alone. We do this in Section 2.3, where we show that this is also true of the NOR gate.

The OR gate

The gate defined in Fig. 9 performs the mathematical operation of inclusive-or, that is 'either *x* or *y* or both'. It only fails to produce a 1 output when both inputs are set at 0.

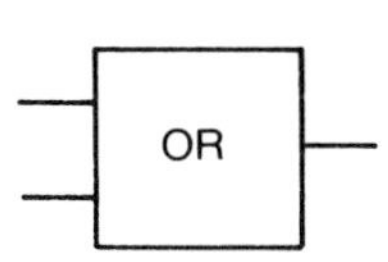

INPUTS	OUTPUT
0 0	0
0 1	1
1 0	1
1 1	1

Fig. 9

The NOR gate

In a similar fashion to the NAND gate, the NOR gate can be seen as a NOT-OR gate. Again, like the NAND gate, it is sufficient in that all gates could be constructed from a suitable combination of NOR gates alone. It is defined in Fig. 10.

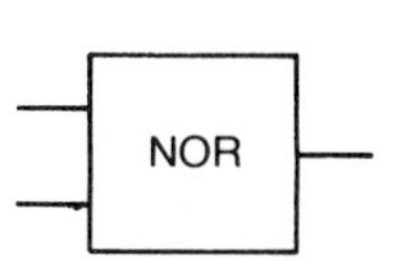

INPUTS	OUTPUT
0 0	1
0 1	0
1 0	0
1 1	0

Fig. 10

The EQUIVALENCE gate

An important gate is defined in Fig. 11. It responds with a 1 output whenever the two inputs are the same, regardless of what they are.

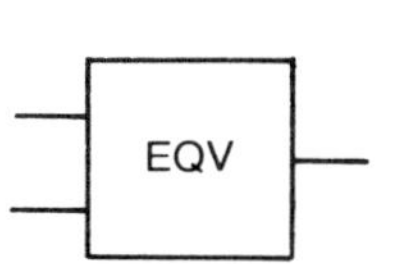

INPUTS	OUTPUTS
0 0	1
0 1	0
1 0	0
1 1	1

Fig. 11

The NON-EQUIVALENCE gate

In the now familiar way, this gate may be considered as a NOT-EQUIVALENT gate, that is an EQUIVALENCE gate followed by a NOT gate. It is also the EXCLUSIVE-OR gate, in that it represents the mathematical operation of 'either x or y, but not both'. It can thus have two symbols, both of which are shown in Fig. 12.

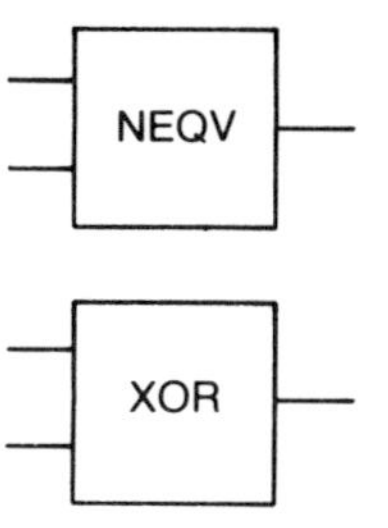

INPUTS	OUTPUT
0 0	0
0 1	1
1 0	1
1 1	0

Fig. 12

2.2 Bistable circuits

We noted earlier that the gate components of Section 2.1 did not allow the 'history' of the gate to influence its behaviour. This is not entirely satisfactory in a binary computer. There is a need for circuits that will 'remember' bits, and that may be switched from one state to another on receipt of the appropriate signals. Such two-state components are known as *bistables*, because they can be stable in either of two states, arbitrarily representing 0 and 1.

Although we have talked in Section 2.1 about bit-signals arriving as pulses, driven in some unspecified way by a clock, understanding the circuits of this section requires us to modify this view in two ways. First, we must consider a 0-bit or a 1-bit as being represented by a sustained signal; secondly, we must show explicitly how the clock fits into our circuits. First of all, however, we consider a simple bistable component with a single output and two inputs that operates on sustained signals and not pulses; we shall then show how to drive this circuit by a clock.

The asynchronous R-S bistable

'Asynchronous' simply means 'without synchronized timing' and the rather formal title 'R-S bistable' simply refers to the two inputs which can be represented as a SET signal and a RESET signal. This component can be considered as a combination of simpler gate circuits as shown in Fig. 13, and it is instructive to do so. Let us suppose that the circuit is initially in the state shown in Fig. 13 with sustained 0-signals on both inputs and a 0 on the output. What happens when the signal on the S (SET) input becomes a 1? The upper NOR gate produces a 0-bit. Thus the input of the lower NOR gate goes to zero, and the output registers 1.

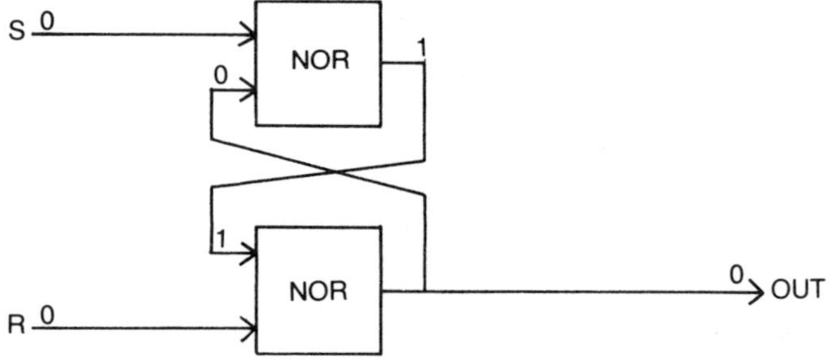

Fig. 13 Asynchronous R-S bistable

More interestingly, however, the output is fed back into the first NOR gate. It is now possible to remove the signal on the S input without the system changing its state. No subsequent change to the S input will have any effect on the system until after a 1-signal is applied to the R (RESET) input. In this case, the lower NOR gate outputs a 0, having the dual effect of changing the output to a 0, and also removing the 'feedback' signal which was holding the upper NOR gate. When sufficient time has elapsed for the system to settle down into its other state (as it was at the start), the RESET signal may be removed without altering the state of the system.

This system of gates thus has two stable states, each of which maintains a continuous signal on the output. The system changes to a 1 on receiving a signal on the SET input, and to a 0 on receiving a signal on the RESET input.

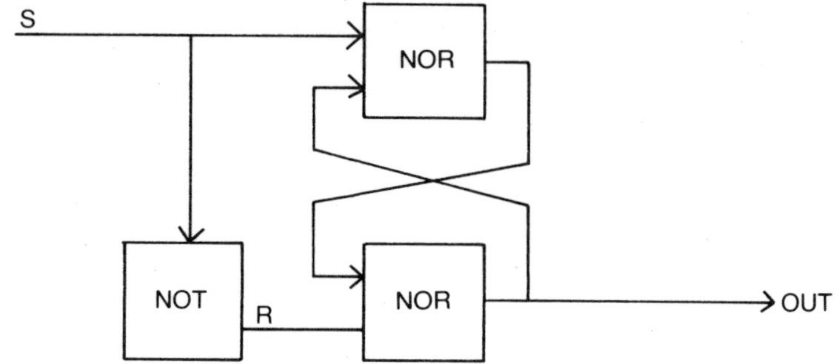

Fig. 14 Modified 1-input bistable

Fig. 14 shows a modified form of this bistable circuit. The original circuit responded only to 1-bits on the appropriate inputs. The addition of the extra NOT gate means that whenever the signal on the S input goes to zero, a 1-bit is fed to the R input. Hence the modified circuit 'remembers' the signal present on the S input. By itself, this is not very remarkable. After all, the system of gates is not doing any more than the S input is achieving by itself without gate circuits! In order to make some constructive use of this modified circuit, we must arrange some synchronization, and this is done in the next section.

The synchronous R-S bistable

We observe in the next chapter (Section 3.4) that electronic circuits take a finite time to settle down to a stable state, and while this is happening, outputs from the circuit may present erroneous values. This problem is avoided by driving the whole computer system by a clock with a carefully chosen clock rate. Outputs are only looked at on the 'tick' of the clock. By design, the clock only ticks at appropriate intervals after all circuits have settled down. We can show how the clock is made to affect the system, by amending our asynchronous R-S bistable.

The SET and RESET inputs are, necessarily, the outputs of some other components. (In our modified bistable, there is essentially only a single input.) We must arrange, by using the clock pulses, that this input is only looked at for the duration of a clock tick. Fig. 15 shows how, by the

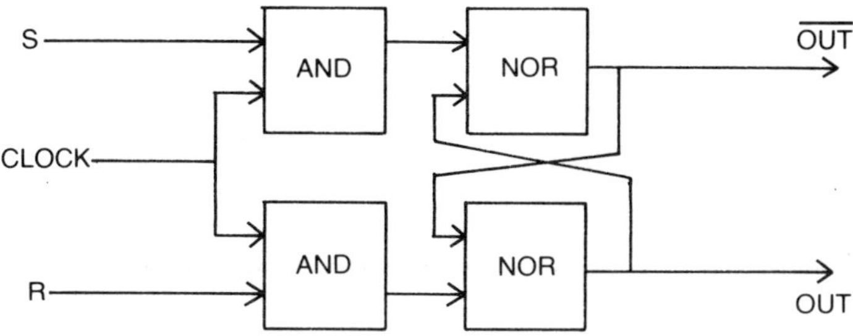

Fig. 15 Synchronous R-S bistable

addition of some AND gates, this may be achieved. Our bistable is now *synchronous* and only capable of changing state during a clock pulse.

We have indicated a further modification to the bistable circuit in Fig. 15. It is left to the reader to check that the additional output $\overline{OUT}$ is always maintained at 0 when the original output is at 1, and vice versa. A circuit (such as the adder/subtractor of Section 3.5) which may require a negated input, can obtain it without the use of further gating circuits by selecting the appropriate one of the OUT and $\overline{OUT}$ outputs. This is particularly so where gates are supplied in standard forms, packaged on a micro-electronic 'chip' with all possible input/output terminals to the gates provided, in order to make a cheap general-purpose component.

The parallel register

Our examples, throughout the book, are of arithmetic operations carried out on 8-bit words. Clearly, 8 synchronous R-S bistables arranged in parallel may be used for storing these 8 bits at any point. It is convenient however, to replace the regular clock input by two separate inputs that hold a 1 (we sometimes say that an input is 'asserted' when it holds a 1-signal) when a read or write operation is required, respectively. These inputs will be asserted as required, but only at clock ticks. To simplify the resulting diagram, we introduce a modified symbol to represent the synchronous circuit of Fig. 15. This is shown in Fig. 16 together with a symbol to represent the synchronous version of the circuit in Fig. 14. Fig. 17 shows an 8-bit parallel register.

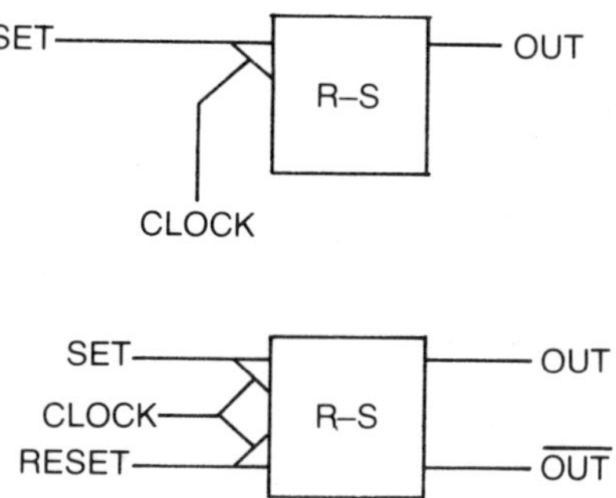

Fig. 16 Symbolic representations of the synchronous circuits of Fig. 14 and Fig. 15 (Notice that the outputs are reversed from those of Fig. 15; the OUT signal is 1 following a SET signal, and the $\overline{OUT}$ is 1 following a RESET.)

The shift register

It is often necessary to shift the bits of a word. This may happen when, for example, the bits are arriving, one following another, from some serial input device such as a teletype. Or again, it may be an operation explicitly requested by the programmer. (See Chapter 3, p. 25 for a discussion of the shifting options that may be called upon by a programmer.) Using synchronous bistables, and in particular the modified 2-output circuit of Fig. 15, such shifting is easy to effect. The output of each bistable is fed to the SET input of the next bistable, and the complementary output to the RESET input. The importance of a well-chosen clock rate, used as a shift-pulse, becomes particularly obvious here. The clock pulse will cause each bistable to take on the state of the previous bistable; the present state will be read simultaneously into the next bistable in sequence. However, the clock pulse must be removed before the output of each bistable takes up its new state, or else the new state will, in turn, be read into the next bistable along the chain.

The bistable shift-register shown in Fig. 18 shifts bits right one position every time a clock pulse of sufficiently short duration is applied to the clock input. It is a relatively simple matter to adjust the design of this register to cause it to shift left, or to shift circularly, feeding the output from one end of the register in again as the input at the far end. Such shift

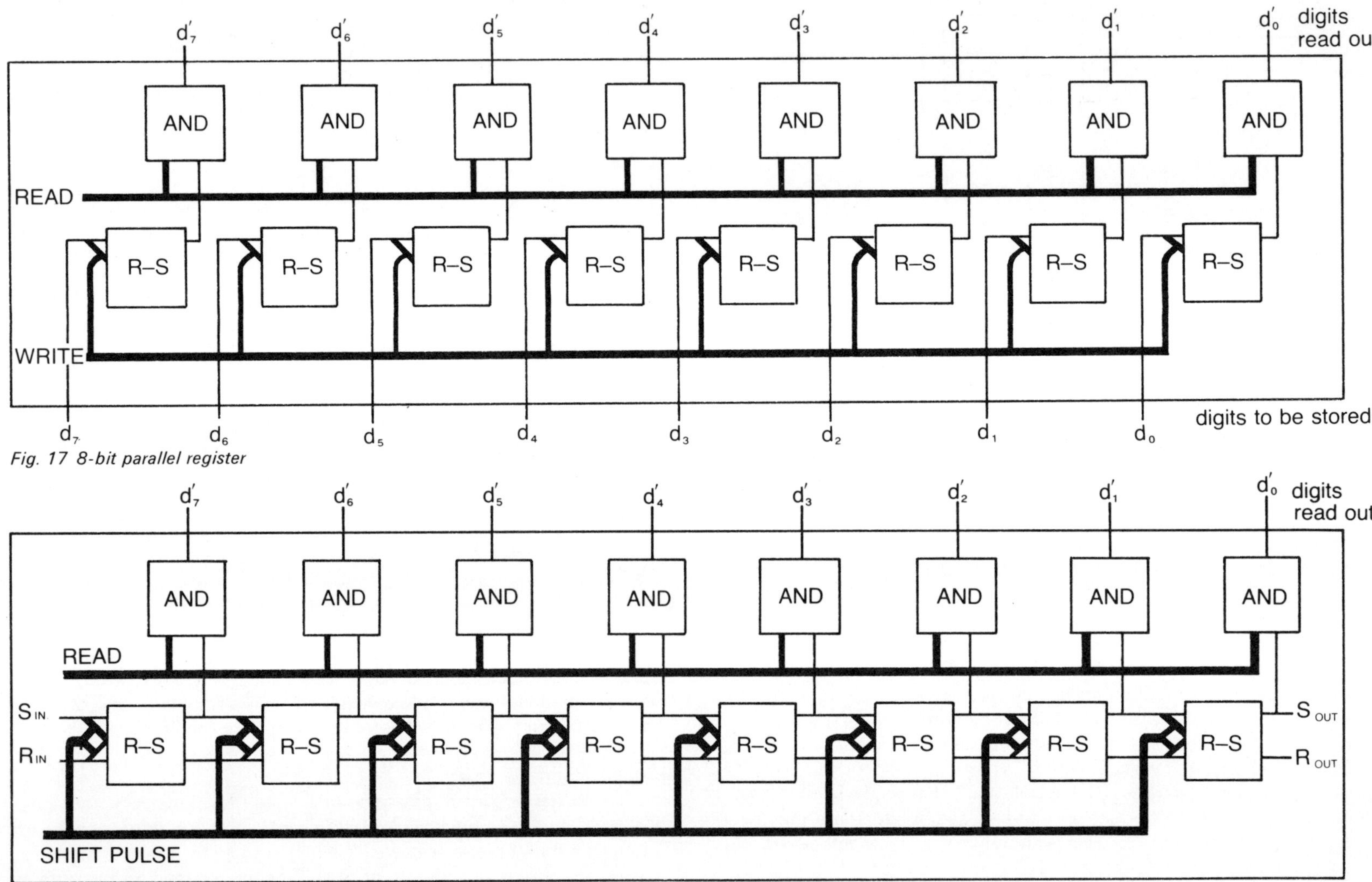

Fig. 17 8-bit parallel register

Fig. 18 Right-shift register

operations are described further in Chapter 3. To read the current contents of the shift register, a pulse is applied to the READ input. As shown, digits can only be entered via 8 shift operations; extending the circuit to provide a WRITE input is a relatively easy matter.

2.3 Equivalent circuits

As we have already noted, the behaviour of any logical component in the computer can be characterized in terms of the outputs resulting from combinations of signals on the inputs. It can often happen that a complex circuit can be seen as a combination of simpler components in more than one way. For example, both the circuits shown in Fig. 19 behave in the same way, although their internal construction is very different. (In fact both are equivalent to the NEQV or XOR gate). A study of *Boolean Algebra* can lead to a profitable reduction in the complexity of circuits, and also to a surprising and unexpected result: that it is possible to construct all logic circuits using just NAND gates (or alternatively, just NOR gates).

Rather than establish this formally, we shall just observe that all the circuits of this chapter (and in fact all circuits) can be constructed from AND, OR and NOT gates alone. This can be justified mathematically. If we can show that these three basic gates can be constructed from NAND gates, it will follow that all circuits can be built using only NAND gates. Fig. 20 does just this, showing circuits equivalent to a NOT gate, an AND gate and an OR gate, each using just NAND gates. The reader is left to construct similar circuits using just NOR gates.

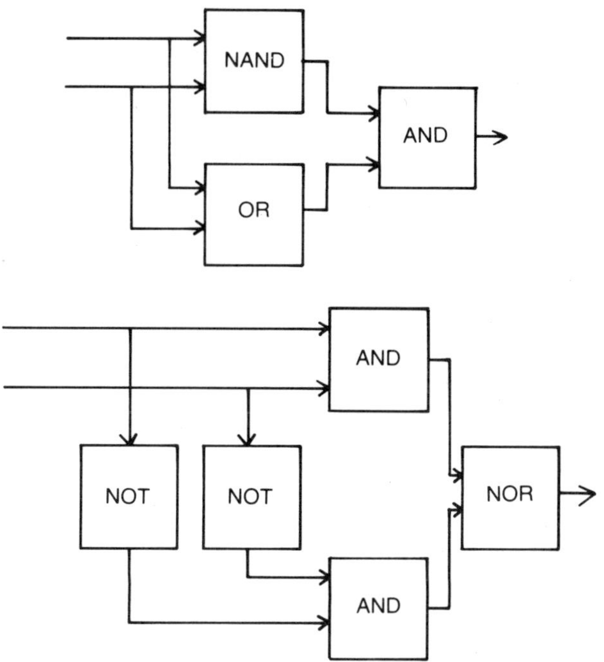

Fig. 19 Two equivalent circuits

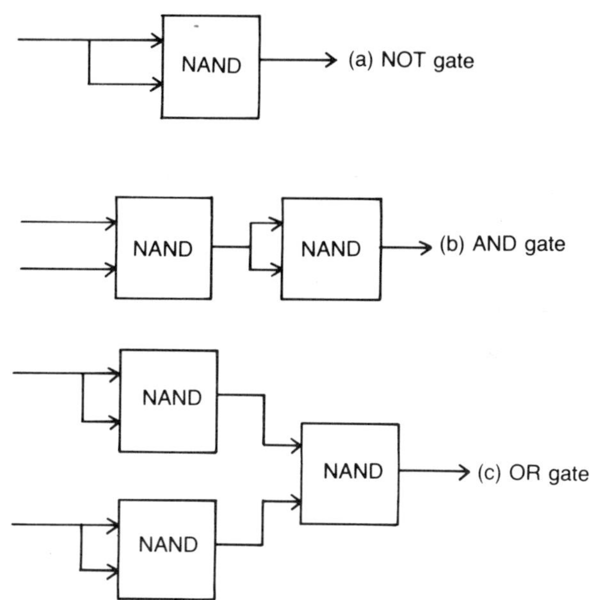

Fig. 20 Gate circuits using only NAND gates

2.4 Some arithmetic circuits

The fundamental arithmetic circuit is the *adder.* Chapter 3 will show that with this circuit as a basis, a wide range of arithmetic processes can be carried out. There are in fact two variants.

The two-input adder (half-adder)

This circuit is a 2-input, 2-output circuit. Fig. 21 shows one way in which it can be realized using gate circuits, and Fig. 22 shows the relationship between the inputs and the outputs. It will be seen that the outputs correspond to the SUM and CARRY of binary addition.

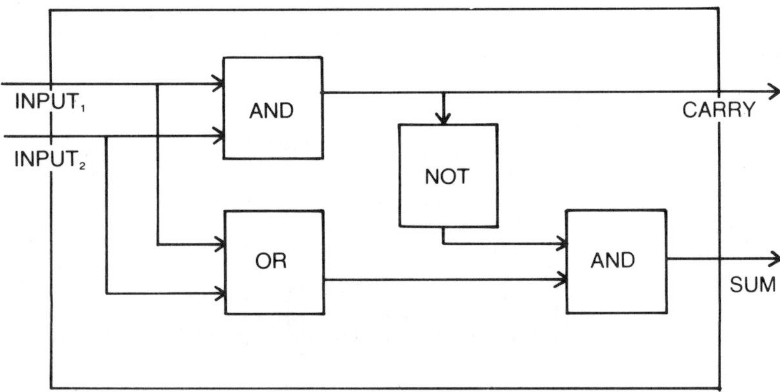

Fig. 21 Two-input adder (half-adder)

INPUTS		SUM	CARRY
0	0	0	0
0	1	1	0
1	0	1	0
1	1	0	1

Fig. 22

Although this is a useful circuit, and behaves as required when adding two bits together, we require a slightly more complex circuit, in order to deal with the addition of carries from earlier adder circuits.

The three-input adder (full-adder)

This circuit adds together three binary inputs, one bit from each number being summed and one carry from the previous stage. The arrangement of these three-input adders to form an adder circuit (in fact, an adder/subtractor) is shown in Chapter 3. The circuit can be built up using two half-adders, as shown in Fig. 23 and its behaviour is shown in Fig. 24.

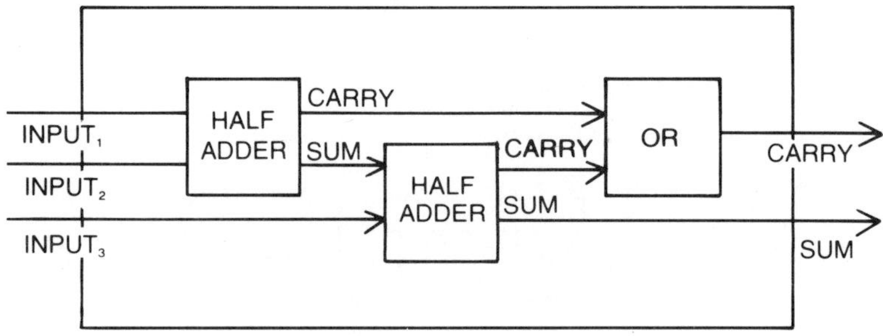

Fig. 23 Three-input adder (full adder)

INPUTS			SUM	CARRY
0	0	0	0	0
0	0	1	1	0
0	1	0	1	0
1	0	0	1	0
0	1	1	0	1
1	0	1	0	1
1	1	0	0	1
1	1	1	1	1

Fig. 24

2.5 Masking

We have already seen that the AND gate can be used in some sense as an 'enabling' gate; thus, in Fig. 15, rather than pass the S and R inputs directly to the bistable, an AND is performed between each of these and the CLOCK input. Only when the CLOCK input is asserted do the values of S and R pass to the bistable. Thus we obtain the synchronous bistable of Section 2.2.

This 'enabling' characteristic of the AND gate can be used to advantage in situations where we are interested in only part of the contents of a computer word. Consider for example a 16-bit/word computer. An instruction word might well look like Fig. 25 (a), and a floating-point word might be divided as in Fig. 25 (b). Characters might be stored using 6 bits/character and be packed into a word as shown in Fig. 25 (c). In each of these cases, we need to have access to only part of a word, temporarily ignoring the remaining information. Thus, in the instruction word, we shall require to separate the operation code, the indirect bit, the address and so on; in the floating-point word, we shall separately require the mantissa and the exponent, and so on.

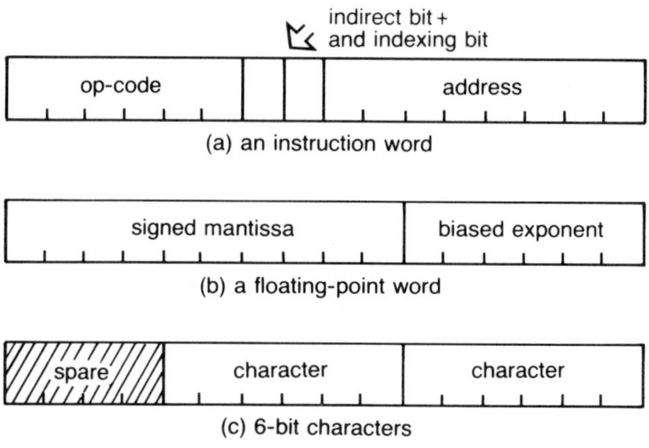

Fig. 25 Possible divisions of a 16-bit word

We can access any particular *field* of a computer word, by setting up a *mask*; this is simply a full-length word with a 1-bit in each position corresponding to a relevant position, and a 0-bit elsewhere. Thus a mask corresponding to the address portion of an instruction word would be the 16-bit word

0000000011111111

and that corresponding to the leftmost character in Fig. 25 (c) would be

0000111111000000

Such masks can be used in the following way: an AND operation is performed on all 16 bits of the mask and the corresponding bits of (say) the instruction word. The result will be to 'blank off' all irrelevant bits of the instruction word, giving a word which has a copy of the required field (in the corresponding position) and 0-bits elsewhere. A suitable shift operation (see Chapter 3) will position the field at the least significant end of the word, which is where it is usually required.

Consider, as an example, the extraction of the 4-bit exponent from the 16-bit floating-point word

	0111001010111010
the mask is	0000000000001111
AND produces	0000000000001010

which is already at the right-hand end of the word, and does not require any shifting. The extraction of individual fields is usually known as *unpacking*. The reverse process, known as *packing*, can be achieved by a combination of shifting and OR instructions. Thus, to pack the two characters 101010 and 111001, initially held in full-length words, into the format of Fig. 25 (c) we have

CHAR1	0000000000101010
CHAR2	0000000000111001
CLEAR ACCUMULATOR	0000000000000000
SHIFT CHAR1 LEFT	0000101010000000
OR WITH ACCUMULATOR	0000101010000000
OR CHAR2 WITH ACC	0000101010111001

2.6 Matching

Much of the power of a stored-program computer comes from its ability

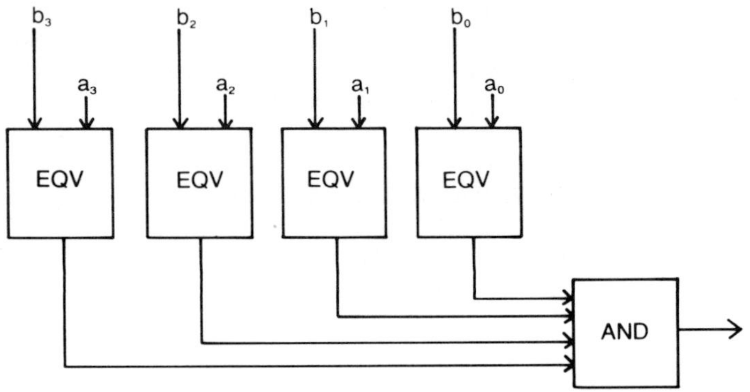

Fig. 26 A matching circuit

to execute conditional statements. Much of this power may be traced back to the EQUIVALENCE and NON-EQUIVALENCE gates, which output signals as the result of a comparison between the individual bits of a word. Fig. 26, for example, shows one possible circuit in which the output is 1 if and only if each of the input bits match: that is, if the two words are identical. Although shown for 4-bit words, the principle clearly extends to words of any size. The circuit uses an AND gate with 4 inputs.

Sometimes, it is more convenient to match words by use of the NEQV gate (also known as the EXCLUSIVE-OR gate). In this case, as shown in Fig. 27, the 4 output bits form a word of all zeros if and only if the input words are equal. The resulting word may be electronically easier to identify or to process.

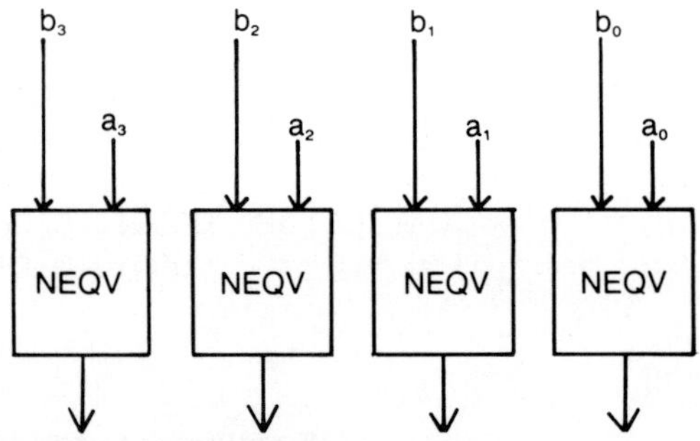

Fig. 27 An alternative matching circuit

While we have observed that more complex circuits can be realized by combining the simpler circuits with two inputs, it is common to find that AND and OR gates in particular are built using multiple inputs. Thus, the 4-input AND gate of Fig. 26 produces a 1-bit on output only when all 4 inputs are each 1. A 4-input OR gate would produce a 1-bit on output if any one or more of the inputs was asserted.

Summary

This chapter has exhibited a number of simple logical circuits, and discussed them solely in terms of the relation between their outputs and their inputs. These circuits are the building blocks of the computer, but the technology used to realize them changes rapidly. At one time, diodes would have been used for the simpler gates; now a large number of gates can be bought cheaply, packaged together on a 'chip' (a micro-electronic printed circuit).

EXERCISES

Public examination questions

1. *A computer logic designer is embarrassed by a shortage of the components necessary to build AND, OR and NOT logic gates. He does, however, have a plentiful and cheap supply of a single logic component, with two inputs* X *and* Y, *and a single output denoted by* X∗Y *and defined by the table*

X	Y	X∗Y
1	1	0
1	0	1
0	1	1
0	0	1

where 1 represents the presence, and 0 the absence of a detectable signal.
Draw similar tables for the compound operations

$$(X*Y)*(X*Y)$$
$$(X*X)*(Y*Y)$$

and identify these operations by giving their commoner names. Define the NOT operation in terms of the ∗ *operation. Indicate how the designer can escape his predicament, by writing expressions for the sum and carry outputs from a half-adder using only the* ∗ *operation.*
(London, 1975) *(15%, 3 hours)*

2. *Given that the three locations* a, b *and* c *contain the patterns*
   a. *10101010*
   b. *11011011*
   c. *10010010*

   evaluate the logical expressions

   $(a \wedge b) \vee c$ and $a \wedge (b \vee c)$.

   ($\wedge$ *= AND,* $\vee$ *= OR)*
   (London, 1976) *(≃4%, 3 hours)*

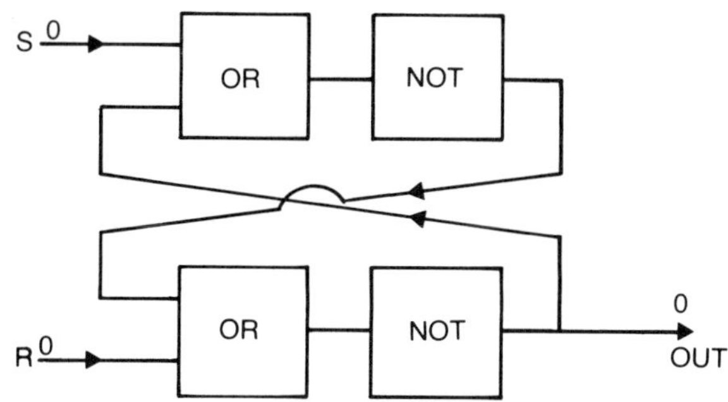

3. *The logic circuit shown in the diagram above is initially in the state shown, with a 0-signal on both inputs and on the single output. Describe the changes, if any, to the output when the following happen in sequence*
   a. *a 1-signal is placed on the* S *input,*
   b. *that signal is restored to zero,*
   c. *the 1-signal is placed on the* R *input,*
   d. *that signal is restored to zero.*

   (London, 1977) *(4%, 3 hours)*

4. a. *For the circuit shown in the diagram below construct a truth table to show, for all possible combinations of input values (1 or 0) at* A, B *and* C*:*
      (i) the values of the inputs D *and* E *to the NOR gate;*
      (ii) the values of the output at F.

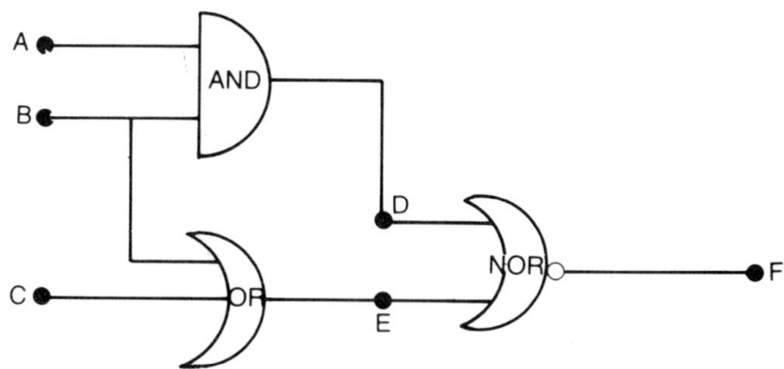

   b. *From your table, state what effect, if any,* A *has on the final output at* F. *Hence find one single gate with only two inputs which would replace this whole circuit.*
   (Oxford) *(7%, 3 hours)*

5. *The truth table for the* nor *operation is as follows:*

p	*q*	*p nor q*
∅	∅	1
∅	1	∅
1	∅	∅
1	1	∅

(∅ represents zero.)

The not operation may be defined in terms of the nor operation, by

not p = p nor p

Define the and operation in terms of the nor and not operations.
(Cambridge Specimen Paper) (≃4%, 3 hours)

Project

6. *Write a program (in any high-level language) to accept a logical expression, check it for syntax errors, and to print out a truth table for the expression. You may like to extend this to provide graphical output.*

3 Fixed-Point Arithmetic

Think of a number. Double it. This simple step is fundamental to the logic of fixed-point arithmetic. We shall see in this chapter that multiplication in particular makes considerable use of shifting or scaling operations. We shall require these operations again in the discussion of floating-point arithmetic.

3.1 Fixed-point shift operations

In order to multiply a denary number by 10, we shift it one place to the left. Thus

1234 ×10	becomes	12340
1.234 ×10	becomes	12.34

and so on. The situation is exactly similar in binary arithmetic

$$1101_2 \times 10_2 \quad \text{becomes} \quad 11010_2$$

Thus, doubling a number in a binary representation corresponds to a left-shift by one place. Multiplying by 4 corresponds to a 2-place left-shift, dividing by 8 a 3-place right-shift, and so on.

This is not quite the whole story, however. We are limited in the amount of shifting possible, by the number of bits in each word, and we must give some thought to what is shifted in to fill up the spaces. There are three different sorts of shift operation possible.

The logical shift

This is the most straightforward: it does not interpret the bit-pattern at all, and merely shifts the bits the required number of places, losing any bits shifted out at either end, and replacing them by 0-bits shifted in at the other end.

For example:

starting with	01011110
1-place logical left-shift	10111100
1-place logical left-shift	01111000
1-place logical left-shift	11110000
and so on.	

In our example, a logical left-shift of eight places would always result in a word consisting of all zeros 00000000. So would a logical right-shift of eight places.

The logical shift only corresponds to doubling or halving while 0-bits are being shifted out. When 1-bits are being shifted out, the logical right shift is approximately halving, because information of low significance is being shifted out. For example:

start with	00001011	(denary 11)
1-place logical right-shift	00000101	(5)

When a logical left-shift loses a 1-bit, significant information is lost.

The logical shifts are of little use with signed representations of numbers, in that they attach no significance to the sign-bit. Thus, with the 2's-complement signed integer 10000111, a logical shift by one place in either direction will change a negative number into a positive one.

The arithmetic shift

The arithmetic shift goes some way to counteracting this. While shifting left, 0-bits are shifted in at the right. However, while shifting right, copies of the sign-bit are shifted in. Thus

starting with	10011001
1-place arithmetic right-shift	11001100
1-place arithmetic right-shift	11100110
1-place arithmetic right-shift	11110011
and so on	

starting with	00011001
1-place arithmetic right-shift	00001100
1-place arithmetic right-shift	00000110
1-place arithmetic right-shift	00000011
and so on	

You may check that right-shifts are still approximately halvings. We are still bound to lose significant information with a left-shift that loses a 1-bit, and we may still turn a negative number into a positive one, as the result of a left-shift. It is worth noting that on some computers the sign-bit does not change during any arithmetic shift.

The cyclic shift

For completeness, we mention a third shift, also known as a *rotation*. This shifts in the bit that has just been shifted out, and hence rotates the bit-pattern. With an 8-bit word, an 8-place right- or left-shift leaves the word unchanged. This operation is of most use in the manipulation of (say) characters packed several to a word, possibly in conjunction with masking operations, as we saw in Chapter 2.

3.2 Overflow

We shall see that fixed-point arithmetic operations involve shifting as part of the algorithms. Almost always this is a left-shift. Where the shifting loses a 1-bit, significant information is being lost. This is known as *overflow*. The term overflow is widely used to describe any situation in which the result of an operation is impossible to store. Thus, it also describes the result of left-shifting the (2's-complement) bit-pattern 01111111 by 1 place. This changes +127 into −2 rather than +254.

An analogous situation, called *underflow*, occurs with fractions or floating-point representations, where all significance is lost and a positive result must be stored as zero.

In high-level languages, overflow and underflow are usually treated as error conditions, and may sometimes cause execution to stop. Alternatively, the result of an operation causing overflow may be set to the largest quantity the computer can store ('infinity' as it were) with or without a warning message. At the machine code level, overflow generally causes an interrupt (see: P. J. Barker, *System Software* p. 37) or sets a flag in the CPU that may be tested by the programmer.

3.3 Double-length registers

It will emerge in what follows, that the processing of fixed-point representations cannot always be carried out in words the same length as storage locations. For the simplest example, consider that the product of two 8-bit words is 16 bits in length. Even where the final product does not cause overflow, the shift operations involved may necessitate the use of a full 16-bit register. It is common, therefore, in the CPU, for operands to be held in double-length registers. The 8 bits in store are found as the right-hand 8 bits of the register, with the left-hand 8 bits either all 0's (for a positive number) or all 1's for a negative number.

This is compatible with the arithmetic shifts discussed above, which can be seen to treat the word being shifted as a section of a much larger register, extended by 1's or 0's as appropriate. The test for overflow is now not simply the loss of a significant bit, but rather the presence of any significant information in the left-hand 8 bits that will be discarded before storing. It is in this way that high-level languages can switch from fixed-point to floating-point representation when fixed-point overflow is detected. Information has not yet been lost—it is still present in the double-length registers. Most BASIC implementations switch from integer print-out to 'exponential' (floating-point) print-out in this way.

3.4 Fixed-point addition

The problem in fixed-point addition is to take two 8-bit fixed-point operands and to produce an 8-bit result (the sum), signalling overflow where this is not possible. We cannot treat the problem exhaustively. There are too many individual choices—which method of representation are we to choose? Are we to perform addition serially (one bit from each operand at a time) or in parallel?

We shall examine the problem of adding two 8-bit signed integers represented in 2's-complement form. Our method will be a parallel one—that is, we shall construct an adder that will accept two 8-bit operands and will produce a complete 8-bit sum as its result, rather than producing the result one bit at a time.

Because we have already noticed that, in 2's-complement arithmetic, subtraction is simply negation followed by addition, we shall automatically cover subtraction.

Our central hardware component will be the full-adder that we met in

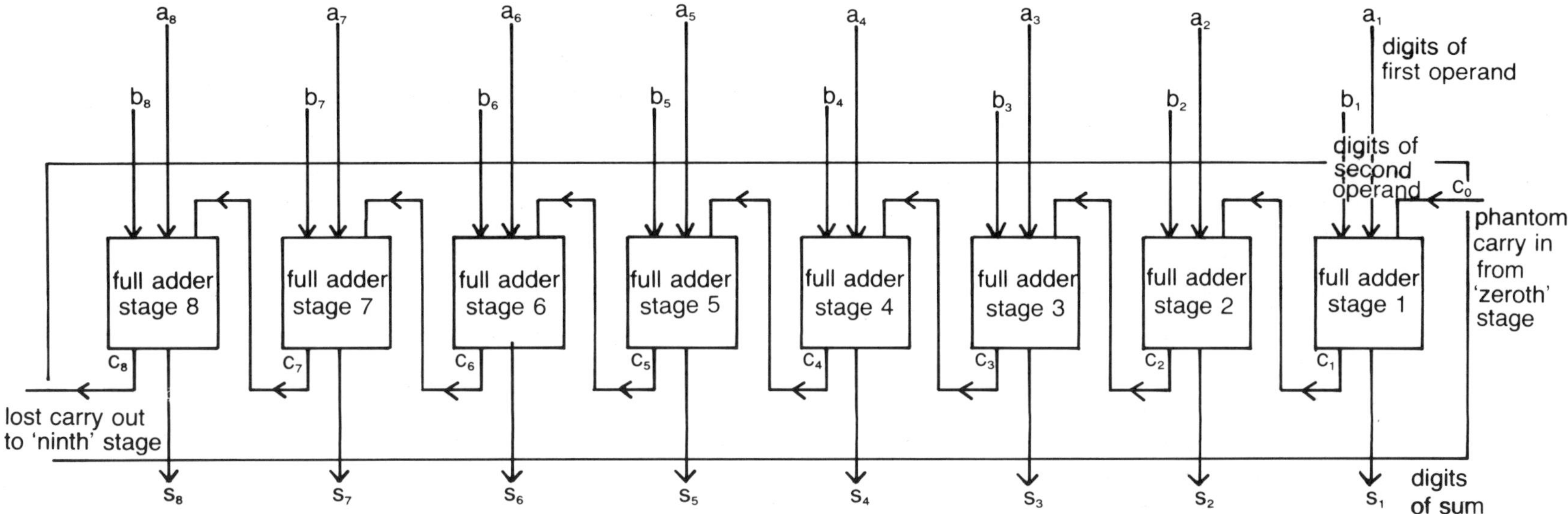

Fig. 28 8-bit parallel adder

Chapter 2 (p. 20). We shall need a full-adder for each pair of digits in the operands, and the resulting arrangement of eight full-adders is shown in Fig. 28.

Notice that the *carry* out of each stage of the adder is an input into the next most significant stage. The carry into the least significant stage, at the right, is not apparently useful. However, it will be worth our while to have it, as we shall see in a moment. For the present, consider it to be permanently fixed to 0.

The time scale of operations in the adder is worth mentioning. The individual full-adders work at an extremely high rate, certainly much faster than the basic cycle-time of the computer. In part, this is necessitated by the arrangements of carries. Initially, all inputs, and hence all carries, will be zero. When the digits of the operands are applied to the inputs each individual full-adder will, after a very small delay, set the 'sum' and 'carry' outputs to the appropriate signals. Each output is set on the basis of a zero-signal on the previous carry. It may happen, however, that this carry becomes a 1-signal after the previous stage has settled down. This will affect the present full-adder, and its outputs will change correspondingly, possibly setting up a new carry into the next stage. This effect will propagate from right to left, and it is not until all the stages have settled down into a steady state that the outputs will all carry their correct signals. This is discussed in Chapter 2.

The problem is common to all electronic components, and it is solved by making the hardware work very rapidly, so that the steady state arrives as soon as possible, and by setting the basic cycle-time of the computer (the clock rate) so that logic circuits such as the adder are only looked at after their outputs have settled into a steady state.

The carry out of the left-hand stage, which represents an attempt to carry into a non-existent ninth stage, is sometimes retained in a special one-bit location where it may be tested by the programmer. We shall not need to use this facility.

3.5 Fixed-point subtraction

Because we have adopted 2's-complement representation, we have a very simple scheme for fixed-point subtraction; this is shown in Fig. 29. It is possible to be a little cleverer than this scheme suggests. It would appear that the operation of forming the 2's-complement of the second operand must be completed before the subsequent addition. In fact, by a minor modification of the adder of Section 3.4, we can turn it into an adder/subtractor.

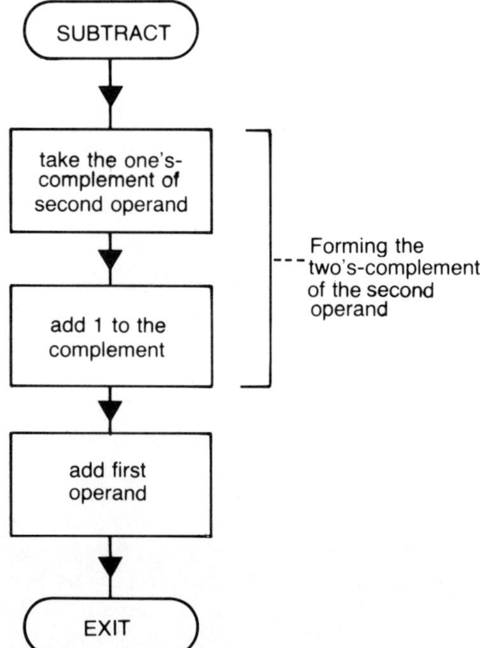

Fig. 29 2's-complement subtraction

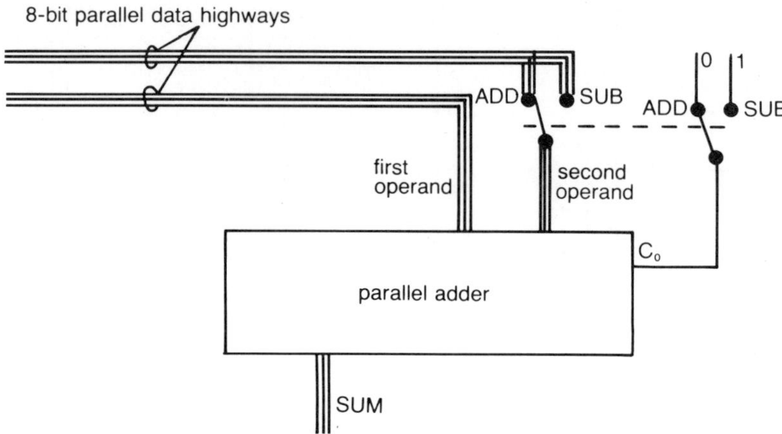

Fig. 30 2's-complement adder/subtractor

Recall that the 2's-complement is always 1 greater than the 1's-complement (for a negative number). Fig. 30 shows how the negation step can be overlapped with the addition step by making use of the 'phantom carry' into the first stage. The digits of the second operand are individually complemented (see Chapter 2, p. 17) thus forming the 1's-complement. The additional 1-bit to be added can be applied during the addition stage.

This modification to the adder turns it into an adder/subtractor. It is not necessary to make use of this modification when adding a negative number, only when subtracting. When adding, the correct answer regardless of sign is guaranteed by the operation of 2's-complement arithmetic.

3.6 Overflow in fixed-point addition and subtraction

The position with fixed-point overflow using 2's-complement addition is rather subtle. We saw in Chapter 1, that ignoring the attempted carry into a 9th bit position was crucial to our understanding of 2's-complement representation. Whenever we add two negative numbers, we shall get a carry from the eighth stage, even where this does not correspond to overflow.

When can overflow occur? Clearly, the sum of a negative and a positive number will be closer to zero than the numerically larger of them, so if both operands are representable within the computer, so will their sum. Overflow can only occur when adding two negative or two positive numbers. Let us look at two situations where this occurs:

11111111	(−1)	01111111	(+127)
+10000000	(−128)	+00000001	(+1)
01111111		10000000	

In each case, overflow is detectable because the sign-bit of the sum is different from the sign-bits of the operands. Notice that in the first example, there is a carry to the 'ninth-stage', while in the second there isn't.

3.7 Fixed-point multiplication

The multiplication of two fixed-point numbers each of 8 bits, to give a product expressible in 8 bits, or else to signal overflow, may be handled in two essentially different ways: repetitive addition or shifting-and-adding.

Repetitive addition

This method is simple, uses the techniques discussed in previous sections, and is almost never found in practice! It uses one of the two factors as a counter, which is decremented while the other factor is repeatedly added to itself. Because fixed-point addition gives an 8-bit answer, so will this method of multiplication. Overflow will be detectable as for addition.

The method is unpopular because it is unduly lengthy, consisting essentially of a loop executed a potentially very large number of times.

Shifting-and-adding

The commonest method of fixed-point multiplication uses the place values of the fixed-point representation, coupled with shift operations. The method is effectively the traditional (denary) algorithm for multiplication, but capitalizes on one of the advantages of the binary system.

In denary multiplication of 353 ×83, the steps are commonly set out as follows:

353	
83 ×	
1059	. . . partial product, 3 ×353
28240	. . . partial product, 80 ×353
29299	. . . sum of partial products

Although the layout above has been labelled accurately, most people see the second partial product as 8 ×353 with an additional 0. This is, of course, a denary shift operation. In computing the significant figures of the partial products, use is made of a 'multiplication table', whether learned by rote or worked out on a scratch pad.

In a binary multiplication, handled in an exactly similar way, the formation of the partial products is even easier. Each product is formed by multiplication by a 0 or by a 1, and hence is either 0 or the original factor. The sum of the partial products is accumulated as the multiplication proceeds.

1011	
101 ×	
1011	. . . first partial product, 1 ×1011
00000	. . . second partial product, 00 ×1011
1011	. . . sum of products (in practice, not computed)
101100	. . . third shifted partial product, 100 ×1011
110111	. . . accumulated sum of partial products

Thus, in the multiplication of two 8-bit fixed-point numbers, one of the factors is left-shifted repeatedly and is added in to the accumulating product whenever there is a 1-bit in the corresponding position in the other factor.

In all, the factor being shifted will be shifted seven times. this must be done in a register large enough to contain the shifted factor at all times, which must thus contain at least 15 bits. The product must similarly be accumulated in a register containing 15 bits, plus a further bit to allow for

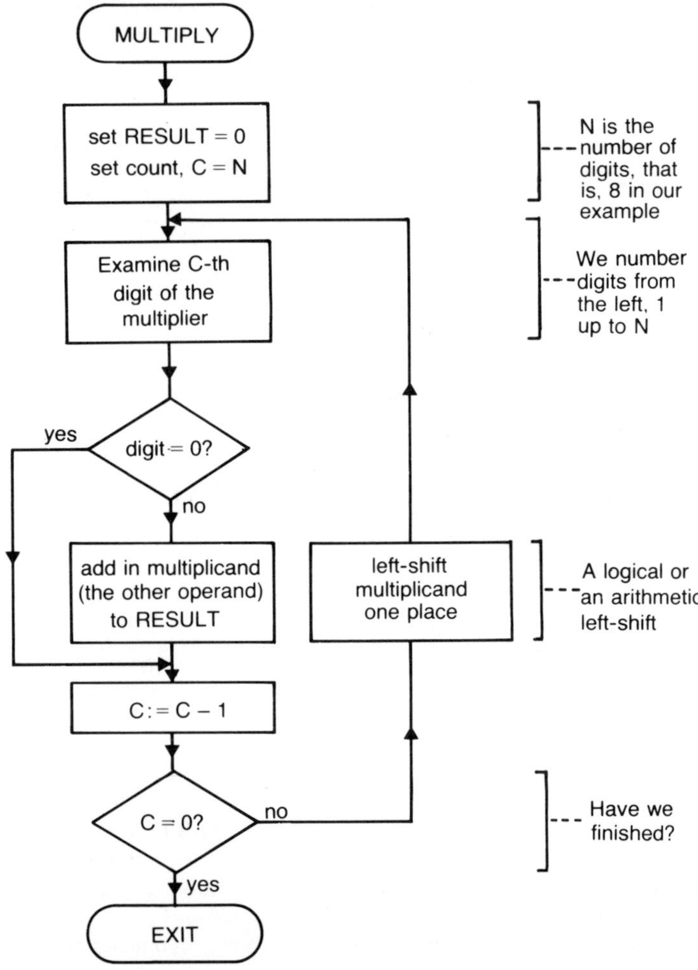

Fig. 31 Fixed-point multiplication

a possible carry to a 16th place as a result of the addition of the last partial product. In practice, two double-length registers are used.

Fig. 31 sets out the algorithm. Notice that the count, C, is decremented from its original value, down to zero. The algorithm works equally satisfactorily with negative numbers. Where one of the complement notations is being employed, the left half of the double-length register must be filled with copies of the sign-bit. And, of course, using 2's-complement notation, we need not actively concern ourself with the algorithm for accumulating the partial sums, as this will simply be regular addition with the signs taking care of themselves.

3.8 Overflow in fixed-point multiplication

The condition of overflow associated with multiplication is essentially a matter of storage. The 16-bit product can always be accumulated without overflow. If the result is to be stored in an 8-bit word, however, the left-hand 8 bits must not contain any significant information: that is, after multiplication, the left half must contain only copies of the sign-bit. In any other case, overflow has resulted and is treated accordingly, possibly with the system software returning a floating-point answer having set a suitable flag.

3.9 Fixed-point division

The division of one 8-bit operand by another to give an 8-bit answer may also be accomplished in two ways paralleling those available for multiplication. There is also a third method, now of little more than historic interest.

Reciprocal multiplication

In the early days of computing, division was a slow process. An iterative method exists for computing the reciprocal of a number (the iterative method does not involve division). It proved to be faster to multiply by the reciprocal of the operand, rather than dividing by the operand. Although both division and multiplication are considerably slower than addition or subtraction (involving as they do, a number of additions), present-day technology regularly uses the subtract-and-shift method described below. Reciprocal multiplication is essentially the equivalent of subtraction by addition of complements.

Repeated substraction

Division is, effectively, repeated subtraction, and may be programmed in that way. As with multiplication by repeated addition, this is a time consuming process, and suffers from the additional disadvantage that division by zero sets up an infinite loop, and must therefore be tested for before the algorithm commences.

Subtracting-and-shifting

This closely parallels the corresponding algorithm for multiplication. It is also close to the system by which division is effected on a hand-cranked mechanical calculator.

The divisor is initially held at the left end of a double-length word. The process is effectively multiplication in reverse, and this is the position in which the multiplier ends up at the completion of the algorithm given in Section 3.7; during subsequent right-shifts, copies of the sign-bit are shifted in (the arithmetic shift) and initially, the double-length word is filled out with zeros.

The other operand (the *dividend*) is held, right-justified in a double-length word. If this operand has come from store (that is, is only 8 bits long) it is filled on the left with copies of the sign-bit. If this is the case, the first subtraction will always fail, but this possibility is not considered in the algorithm, which is set out in Fig. 32.

The algorithm is similar to that for multiplication, but is in some sense reversed. It builds up the quotient, bit by bit, setting the bit currently under consideration to a 1 if the subtraction just attempted is 'possible'. That is, we examine the sign-bits of the dividend before and after the subtraction. The dividend stores the current partial remainder (initially, the value of the dividend, finally the value of the remainder). If the sign changes, the subtraction is 'not possible' and the previous value of the dividend register is restored by adding-in the divisor again. For the reader familiar with the hand-cranked calculator, this is the exact analogy of a

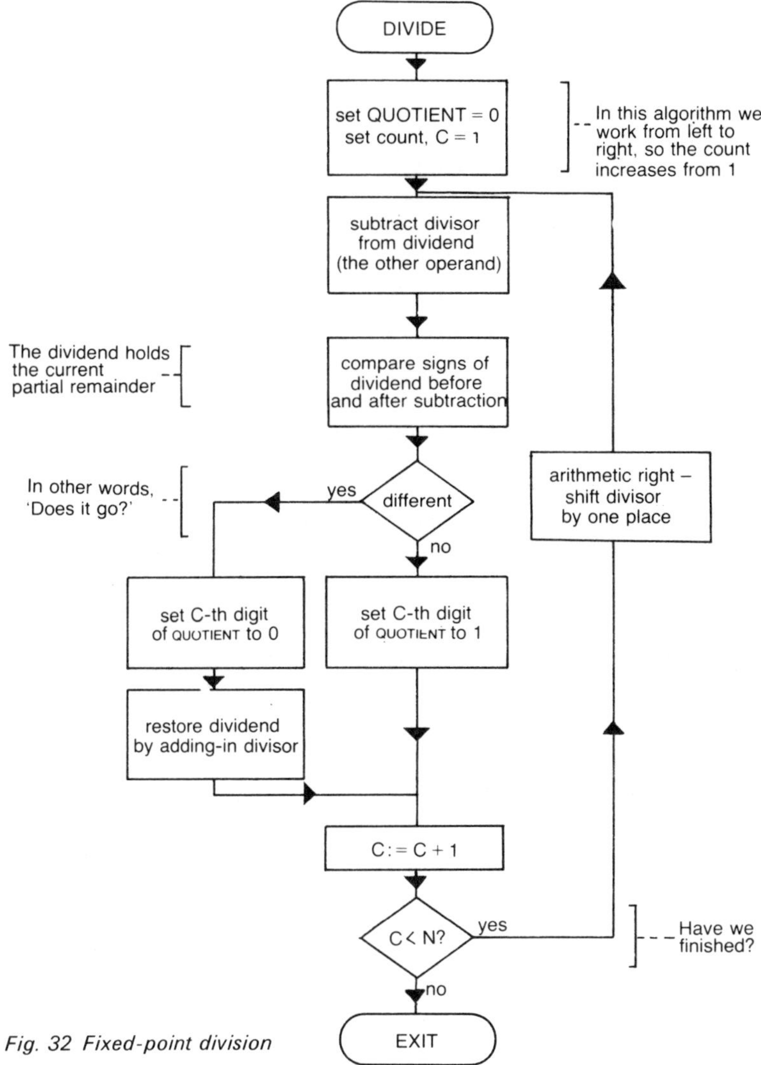

Fig. 32 Fixed-point division

subtracting turn during division that rings the bell, requiring the restoration of the dividend by an addition (a turn in the opposite direction) followed by a shift.

Notice that our algorithm describes integer division (so called even when executed on fractions). This produces a quotient and a remainder. (There is material on integer division in Section 5.8.) Zero division produces a quotient consisting of all 1's, and a remainder equal to the dividend.

3.10 Overflow in fixed-point division

Overflow is detected in division when the quotient is to be stored. The algorithm given in Fig. 32 will always produce a 16-bit quotient and a 16-bit remainder. As we have indicated, the remainder is to be discarded; the quotient is to be stored, and overflow results where the left half (the most significant 8 bits) of the quotient contains significant information. That is, unless the sign-bit of the truncated 8-bit right half (bit 9, numbering from 1 at the left) is the same as all the first 8 bits. Division overflow is treated in much the same manner as other kinds of overflow.

Summary

This chapter has presented algorithms for the four basic operations used in arithmetic: addition, subtraction, multiplication and division, insofar as they can be carried out on fixed-point operands.

Any arithmetic process may result in overflow. For addition and subtraction, which are carried out in single-length registers, overflow is detected as a 'wrong' answer. In the cases of multiplication and division, effected using double-length registers, the full (16-bit) answer is always correct but overflow results when the 8-bit single-length equivalent is not mathematically equivalent—that is, where significant information is present in the left half of the double-length register.

EXERCISES

Chapter material

1. What is the result of a. an arithmetic shift
b. a logical shift
c. a cyclic shift

on each of the following words:

(i) 10000000 (3 places right)
(ii) 01111111 (3 places left)
(iii) 11111111 (8 places left)
(iv) 11111111 (8 places right)
(v) 00011110 (1 place right)

2. Perform 8-bit fixed-point addition in each of the following cases. Which give overflow?

(i) 11111111 + 11111111   *(ii)* 10000011 + 01101110   *(iii)* 10000000 + 11111100

(iv) 01111111 + 00000001   *(v)* 00000001 + 11111111

Public examination question

3. Using a low level language with which you are familiar describe a set of instructions to enable the contents of an accumulator to be shifted. *Include in your answer reference to cyclic, logical and arithmetic shifts. Illustrate in detail the results of shifting the binary representations of +12 and −19 using both left and right arithmetic shift instructions. (London, 1977) (≃ 8%, 3 hours)*

Project

4. Prepare programs to print out information to help you trace the flowcharts in Fig. 31 and Fig. 32.

4 Floating-Point Arithmetic

It is in this chapter that the seeds we have sown in the previous chapters bear fruit. In Chapter 3, we saw how fixed-point representations could be manipulated, and in this chapter we shall discover how to combine the fixed-point algorithms in order to give us algorithms for the addition, subtraction, multiplication and division of floating-point representations.

While fixed-point algorithms are (almost) invariably part of the machine hardware, this is by no means the case for floating-point algorithms. Although floating-point hardware is common, so also is software, either in the form of library routines, or as machine code stored in read-only memory.

The algorithms given in this chapter use floating-point operands to give us floating-point results. As we saw in Section 1.3, it is customary to normalize floating-point representations: Section 4.4 discusses an algorithm for normalizing any floating-point representation. The algorithm is necessary, because the operations of addition, multiplication, etc., may well give unnormalized results, even when applied to two normalized operands.

4.1 Floating-point multiplication

Multiplication is the simplest of the floating-point operations. It is a direct mathematical consequence of the fact that, with two operands each of the form

$$(\text{MANTISSA}) \times 2^{(\text{EXPONENT})}$$

their product is given by

$$(M_1 \times M_2) \times 2^{(E_1 + E_2)}$$

It is only necessary to perform (fixed-point) multiplication on the mantissae, and (fixed-point) addition on the exponents. The exponents may well be stored in excess-n form (see page 8) in which case it will be necessary to adjust for this in the addition.

The algorithm is shown in flowchart form in Fig. 33. It performs satisfactorily when either (or both) of the operands is zero, but it may be more economical to test for this case separately.

If the operands are both normalized, how far from being normalized can the result be? Normalization depends on the form of the mantissa, as the exponent can always be adjusted to compensate, and so we should look at the behaviour of the mantissa under multiplication. With the 8

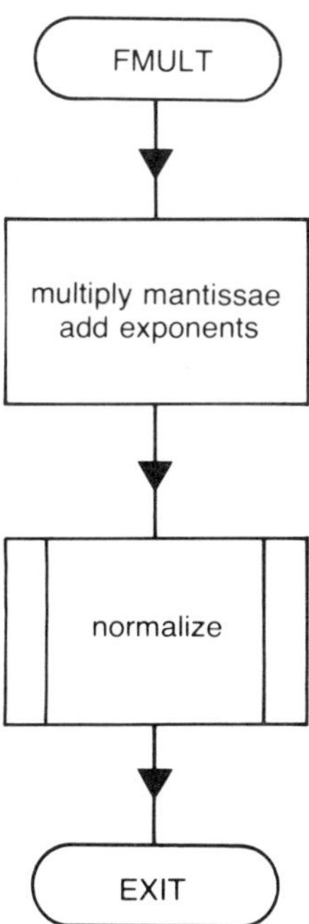

Fig. 33 Floating-point multiplication

bits/word example used in previous chapters, the smallest (normalized) positive mantissa is

0 1000000

which represents 0.5. Hence the smallest product is

0 1000000 × 0 1000000 = 0 0100000

This is not normalized, but a single left-shift (and compensating adjustment of the exponent) will normalize it.
The largest mantissa is

0 1111111

which represents 0.9921875. The largest product is thus

0 1111111 × 0 1111111 = 0 1111110 (00000010)

where the bracketed bits indicate the low-significance part of the product. This product is already normalized, and needs no adjustment.

It is not at all clear that we need to retain these 8 extra bits. Certainly the single 1-bit (which is 2^{-14}) cannot affect the product when it is fitted into an 8-bit word. However, we have seen that some multiplications need a left-shift to normalize the product; in this case we could shift information from these 8 bits into the 'high order' bits. It is simpler to work with complete words, so an 8-bit register will be used to hold this 'low order' information during processing.

Putting these results together, we conclude that a floating-point product is either already normalized, or requires at most a single left-shift to normalize it. This observation will be of some use in Section 4.4

4.2 Floating-point division

This is almost as simple as multiplication, and has a very similar flowchart (Fig. 34). The algorithm is a consequence of the fact that

$$(M_1 \times 2^{E_1})/(M_2 \times 2^{E_2}) = (M_1/M_2) \times 2^{(E_1-E_2)}$$

Again, it is necessary to adjust for the fact that the exponents are probably stored in excess-n form, and it may well be worth testing for a zero dividend, rather than executing the whole algorithm to give a zero result. A zero divisor will show up as overflow, but it may be equally desirable to test for this as well.

With normalized operands, how normalized can the quotient be? Excluding the complications of zero, the largest (positive) quotient occurs with the largest dividend mantissa:

0 1111111

and the smallest (normalized) divisor mantissa:

0 1000000

(Again, we can leave the exponents to look after themselves!) Dividing by $\frac{1}{2}$ is, of course, a doubling, which amounts to a left-shift. Thus, the quotient mantissa in this case is

(1)
0 1111110

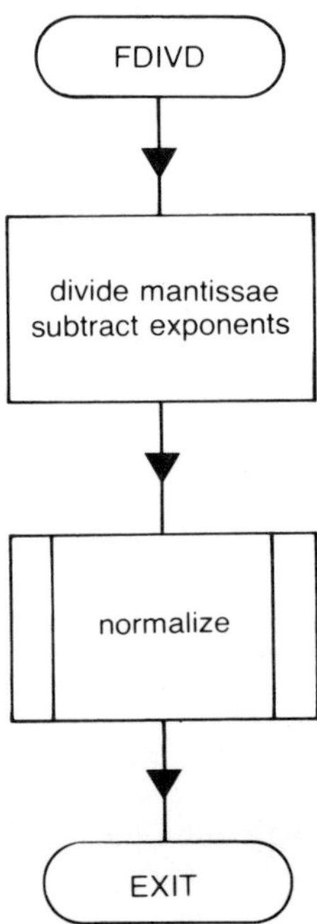

Fig. 34 Floating-point division

where our notation indicates that a 1-bit has been recorded shifting out of the mantissa (we need to preserve the sign-bit intact). This may be detected by the change of sign-bit, in practice, but it is confusing to suggest that we allow this shift to change the number into a negative one. Hence, in this case, a single right-shift will normalize the resulting quotient, together with an adjustment to the exponent.

As this quotient is the largest possible, and as the idea of a right-shift is untidy, an interesting trick is often used to avoid the problem. If the dividend is unnormalized by shifting the mantissa one place to the right before entering the algorithm (and adjusting the exponent), this 'overflow' cannot occur: in the example above, the mantissa becomes 0 0111111 (with one bit shifted into the low order word, if a double-length register is in use) and the quotient is 0 1111111. Thus, in the worst possible case, the result will be already normalized.

At the other extreme, with the division

0 1000000 / 0 1111111

we observe that this becomes

0 0100000 / 0 1111111

The divisor is certainly greater than $\frac{1}{2}$, so the effect is less than a doubling. Only if it were a doubling would the quotient mantissa have leading bit = 1, so we conclude that it requires exactly one left-shift to normalize it.

Putting these observations together, as with multiplication, we conclude that a quotient of normalized operands requires at most one left-shift (and possibly none) to normalize it. This requires that we make use of the trick discussed. Equivalently, without this device, the quotient requires either a single right-shift, or is already normalized.

4.3 Floating-point addition and subtraction

Addition is not as simple as the algorithms of the previous sections. By way of compensation, a single algorithm suffices, as subtraction can be seen as simply the complementation of one mantissa, followed by addition.

The principle is best illustrated by a denary example. We may not directly add together

3.14159×10^6 and 2.71828×10^4

but in the (coincidental) case where the exponents are equal:

3.14159×10^5 and 2.71828×10^5

we can directly add the mantissae, giving 5.85987×10^5. Note that the exponent remains the same, unless the sum of the mantissae is greater than 10, and we wish to 'restandardize' the result.

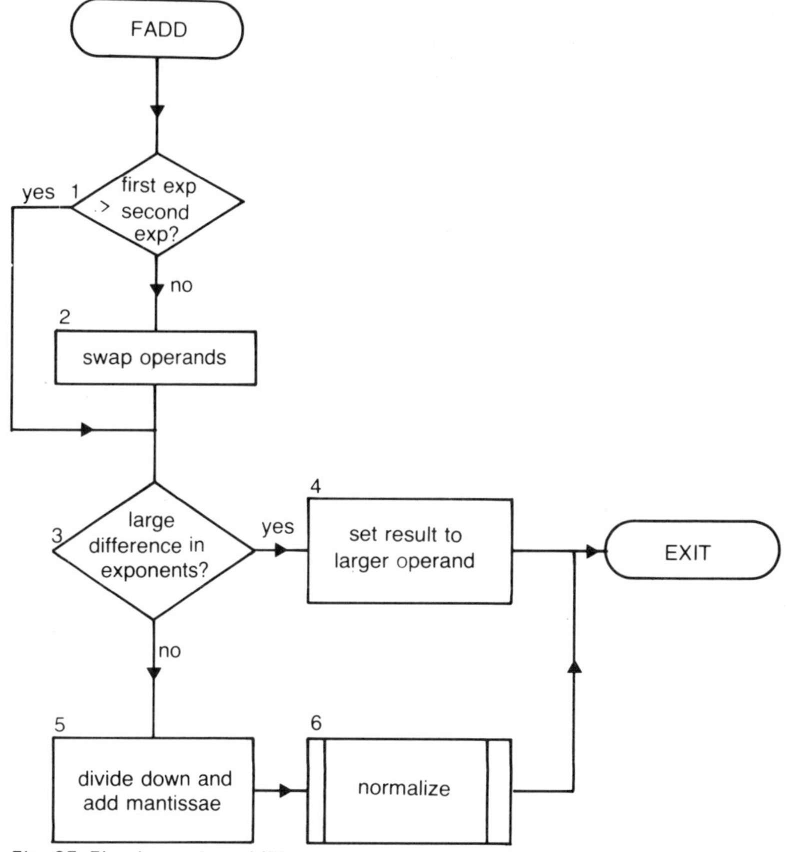

Fig. 35 Floating-point addition

Our original addition could be effected by adjusting one of the numbers to make its exponent equal to that of the other operand:

$$3.14159 \times 10^6 + 0.0271828 \times 10^6$$

whereupon we can add the mantissae, giving 3.1687728×10^6

Floating-point addition in any base can be tackled in the same way. We adjust the operand with the smaller exponent, unnormalizing it in the process, and then use (fixed-point) addition on the mantissae.

There is a danger, with fixed-length words, that the adjustment will shift one operand so much that all its digits (bits) are shifted out. The algorithm, which is flowcharted in Fig. 35, accounts for this. It is worth examining the steps of the algorithm in some detail.

An addition algorithm

Step 1: Test for larger exponent

This test together with Step 2, ensures that the first operand has the larger exponent, interchanging the operands if necessary. This simplifies what follows.

Step 3: Test exponents

If the exponents differ greatly, the process of unnormalizing one, by shifting and adjusting exponents, will lose all significant information. In this case, the effect of the algorithm will be to add what is effectively zero to the larger operand. We can be efficient now, by exiting at once if this is the case, setting the result to the (already normalized) larger operand.

Step 5: Divide down and add

This is the 'unnormalizing' step. We divide down—shift right, by a number of places equivalent to the difference in exponents. We can effect this without considering the excess-n storage of exponents. Nor do we need to adjust the exponent of the unnormalized operand. The purpose of this step is to make it equal to the other exponent, which we know to be the exponent of the answer. We conclude this step by using (fixed-point) addition on the mantissae.

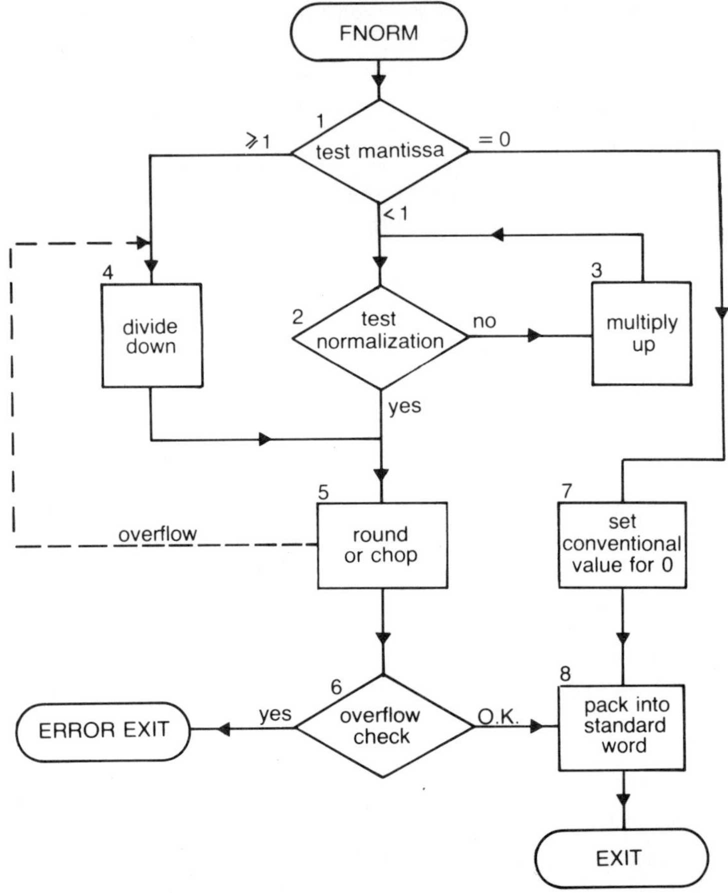

Fig. 36 Floating-point normalization

It is worth being precise about the degree of difference tolerated in Step 3. If the exponents differ by 7 or more, a 7-place shift is required. This will lose all bits from the word being shifted. In general, the test in Step 3 is for a difference of (n −1) where n is the word-length allocated to the mantissa (excluding the sign-bit).

With operands of the same sign (both negative or both positive), floating-point addition could well give a normalized result. At worst, we are adding 'zero' to an already normalized operand, at best we are doubling (and the consequent right-shift can be avoided in the manner of Section 4.2). But the addition of operands of different signs, particularly where they are numerically close, can cause problems of an unusual form which are dealt with more fully in Section 4.6.

4.4 Floating-point normalization

We have seen that each of the algorithms above can be implemented in such a manner that the result requires only left-shifts of the mantissa in order to normalize. Or equally, without 'trickery', at most one right-shift is required. Certainly, a normalization algorithm can be a fairly simple affair.

Fig. 36, which is discussed in detail below, is a flowchart for normalizing any floating-point representation. It would be possible to implement it without allowing for right-shifts, by using the tricks mentioned above.

A normalization algorithm

Step 1: Test mantissa and branch
We first inspect the mantissa: it is convenient to work with the absolute value of the mantissa. This value may be:

a. greater than or equal to 1—needing right-shifts to normalize (branch to Step 4); or
b. less than 1—needing left-shifts or no shifting at all (branch to Step 2); or
c. zero in which case special action is taken at Step 7.

Step 2: Test normalization
If the bit to the right of the 'point' differs from the sign-bit, the quantity is now normalized, and we exit from the algorithm via Step 5; otherwise, we proceed to Step 3.

Step 3: Multiply up
The number is not normalized: it requires at least one left-shift. We execute this shift, with a corresponding decrease in the exponent, and loop back to Step 2. This loop through Step 3 will be executed as many times as necessary to normalize the number. From our earlier comments, this may be as little as once. Eventual success is guaranteed, as the mantissa is not zero, since this has been attended to at Step 1.

Step 4: Divide down
From the previous discussions, it is clear that if any right-shifting is required, it will consist of exactly one shift, so we may execute the shift, compensate with the exponent, and proceed directly to Step 5. This step could be avoided altogether, as has been made clear above.

Step 5: Round or chop
This is a subtle step, in that it introduces an unexpected complication. We are probably working with double-length registers, and we need to produce a single-length mantissa. If we simply truncate ('chop' the extra bits) then we can proceed to Step 6 but if we round, there is a slim possibility that the mantissa rounds up to 1, and we may have to repeat Step 4.

Step 6: Overflow check
Successive alterations to the exponent may have placed it outside the range representable in a single-length register (or whatever space is reserved for exponents); if the exponent is too large, this is overflow, if too small, underflow. In any event, it is a fatal error. Example 4.5 (2) is instructive, in that it shows how the method adopted at Step 5 can affect this test.

Step 7: Zero
Zero cannot be normalized. The conventional representation is usually zero mantissa and exponent.

Step 8: Exit
Our result may now be placed into store in the approved fashion, as two separate words, or as fields packed into a single, longer word.

We may assume that normalization is the last process carried out after all floating-point operations.

4.5 Examples of normalization

Example 1

Let us suppose that we are presented with two double-length registers as follows:

0 0010101 (01001110) (00000000) 00001011

We assume that all 8 bits of the extra capacity carry significant information—that is, the sign-bit of the extra words are not used, except as extra bits. Notice that the 8-bit extension is to the right of the mantissa, but to the left of the exponent. With this double precision, the word above evaluates to 0.16644287×2^{11} or 340.875 (approximately). The precision is too great to store however, and the steps detailed below should be followed with reference to the flowchart in Fig. 36.

Step 1: mantissa is less than 1
Step 2: NOT normalized
Step 3:
0 0101010 (10011100) (00000000) 00001010
Step 2: NOT normalized
Step 3:
0 1010101 (00111000) (00000000) 00001001
Step 2: NORMALIZED
Step 5: chop mantissa to 0 1010101 (or round to same value)
Step 6: exponent becomes 00001001
Step 8: store normalized result as 0 1010101 00001001

This result is 0.6640625×2^9 or 340.0 (approximately). Normalization has introduced an error, in this case of about $\frac{1}{4}$%.

Example 2

Consider two double-length registers

0 0111111 (11111111) (00000000) 11111111

This example is doubly instructive: it underlines the general point about floating-point representations, that the price of inaccuracy at large exponent values is greater than for small values; it also contrasts the results of rounding v. chopping in Step 5. Before normalization, the quantity represented is $0.4999847413 \times 2^{127}$ or (about) 8.5068×10^{37}

Step 1: mantissa is less than 1
Step 2: NOT normalized
Step 3:
0 1111111 (11111110) (00000001) 00000000
Step 2: NORMALIZED
Step 5: chop mantissa to 0 1111111
OR round to (1)
0 0000000

if chopping—Step 6: EXPONENT OVERFLOW—ERROR EXIT
if rounding—Step 4: 0 1000000 (00000000) (00000000) 11111111
Step 5: round mantissa (again) to same value 0 1000000
Step 6: exponent becomes 11111111
Step 8: store normalized result as 0 1000000 11111111

This result is 0.5×2^{127} or approximately 8.5071×10^{37}. The error in the mantissa is 0.0003, but the error overall is 3.0×10^{37}

Example 3

This illustrates both the operation on a negative number, and the introduction of underflow.

1 1101000 (00000000) (00000000) 00000001

Step 1: mantissa (absolute value) less than 1
Step 2: NOT normalized
Step 3:
1 1010000 (00000000) (00000000) 00000000

Step 2: NOT normalized
Step 3:
1 0100000 (00000000) (11111111) 11111111
Step 2: NORMALIZED
Step 5: chop (or round) mantissa to 1 0100000
Step 6: EXPONENT UNDERFLOW—ERROR EXIT

4.6 Floating-point overflow

Because of the way in which we interpret the mantissa, as a fractional quantity, and because we operate with normalized operands, the sources of error in a floating-point result are virtually eliminated as far as the mantissa is concerned. Only division by zero results in a mantissa that cannot be satisfactorily stored, possibly after rounding or truncation.

The principal source of overflow (and underflow) in a floating-point algorithm comes from the exponent going out of range, as in Examples 2 and 3 above.

Potentially more damaging is the error that may be introduced in addition or subtraction. Consider, in a denary example, what will happen if we add together quantities of opposite sign, but approximately equal magnitude (or equivalently, subtraction of nearly equal operands):

$$\begin{array}{l} 0.342437 \times 10^{22} \\ \underline{0.342436 \times 10^{22}} \quad \text{(subtract)} \\ 0.000001 \times 10^{22} \end{array}$$

Up to this point, we have introduced no error, and there is a school of thought that holds that we do no damage by normalizing this result to

$$0.100000 \times 10^{17}$$

However, we would do well to view this result with the utmost suspicion: it implies six (decimal) digits-worth of accuracy, whereas in fact the result is accurate only in the first decimal place. In a chain of similar calculations, it is only too easy to arrive at a final result that is accurate to no decimal places at all!

Beyond noting that we should avoid such a step (if this is possible) by rearranging the order of operations, there is little that we can do about this, except shake our heads in sorrow.

Summary

This chapter has, in one sense, brought us to the end of the road: we now have an understanding of how arithmetic may be accomplished in the computer. But addition, subtraction, multiplication and division are in their turn only building blocks for more complex manipulations: these manipulations are the subject of what follows.

EXERCISES

Public examination questions

1. *In a particular computer, the binary floating point representation of a number*

$$a \times 2^b$$

is stored in two consecutive locations of storage. The first location of the pair contains a *and the second* b, *each held in fixed point form as follows:*

0	1	2	3	4	5	6	7	8	9	10	11

Bit 0 represents the sign of the word (0 for a positive number, 1 for a negative number) and bits 1-11 the magnitude, with an assumed binary point between bits 0 and 1 for a, *and in the second location after bit 11 for* b. *To maintain the maximum amount of information the number is stored in 'normalized' form, i.e. bit 1 of the first word (containing* a*) is 1 unless* a *is zero in which case both words contain all zeros.*

Give algorithms (either as a flowchart or a simple set of instructions) for (a) the multiplication and (b) the addition of two such numbers, given that

(i) the numbers to be operated on are to be found in locations addressed as m, m +1 *and* n, n +1,

(ii) the result will be stored, similarly normalized, in locations addressed as p, p +1.

To simplify your algorithms, you may assume that over-flow and under-flow will not occur and need not be considered.

(London, 1975) *(15%, 3 hours)*

Project

2. *Prepare programs to help you to trace the algorithms of Figs. 33, 34, 35 and 36. Include a switch value, so that if the user requests, the programs of Chapter 3, Exercise 5 may be called to trace intermediate fixed-point calculations.*

5 Commonplace Functions

Think of a number. Take its square root. We cannot easily do this. Nor do we find it easy to take the sine of an angle, the integer part of a number or the logarithm of a number X. But the computer does each of these, with as little difficulty as the addition of two numbers.

This chapter is about ways of computing the functions mentioned above, and some similarly useful ones. There is an important word of warning that must be given at this point. When functions such as these are implemented in a digital computer, they tend to be programmed in a way that capitalizes on the idiosyncracies of the particular machine. The STORE operation might have a useful side-effect, or a constant left over from the system random-number generator might be exactly what is needed for the expansion of the cosine function, for example. What we present in this chapter is a collection of techniques that could be used: it is doubtful if any one computer uses exactly the techniques described here for all its functions. How do we arrive at these techniques? There are two basic methods, described in Section 5.1, and a number of tricks that will be revealed as we go along.

The functions described here are those that are available as part of a language such as BASIC or FORTRAN. There is a list of the available functions as part of the documentation for any language translator. Because they are usually called into action by some form of syntax compatible with the rest of the language, such as

```
LET Y = SQR(X)
```

and thus appear to be 'built-in' to the language, they are often called *intrinsic* functions.

It can be instructive to examine a source code listing of the algorithms used on any particular computer. Sometimes they are part of the operating system, sometimes there is a special library loaded as needed with a compiler, and sometimes they are subroutines forming the main body of the translator (such as a FORTRAN compiler or a BASIC interpreter). Often each language will have its own set of routines, duplicating the work carried out already by authors of other translators. It is, in any case, unlikely that a source code listing exists in the computer. Usually it is the object code only that is stored. It can be perplexing as well as instructive to read through such a set of routines—the short-cuts and idiosyncracies referred to earlier can make it a very confusing experience!

5.1 Methods of evaluating mathematical functions

Mathematicians have provided us with two valuable tools of great practical use in computing values of functions. Both the techniques—that of *iteration*, and that of *power series*, implicitly recognize that many function values cannot be computed to absolute accuracy, even with the most powerful computer in the world. For example, the square root of 2 is 1.4142136. . . . Neither technique works by computing the value $\sqrt{2}$ exactly and rounding it off to fit into a finite length word. Both techniques work throughout with standard length words, gradually getting closer to an answer as accurate as the word-length allows. The treatment here is necessarily sketchy, and suggested references are to be found at the end of the book.

Iteration

Without knowing exactly what $\sqrt{2}$ is numerically, we can see that algebraically it is a value that satisfies the equation

$$x = \tfrac{1}{2}\left(x + \frac{2}{x}\right)$$

You will not be asked to construct such equations for yourself, only to understand how they work. This particular one is a rearrangement of the equation $x^2 = 2$, which is, after all, the equation we are trying to solve. It has been obtained by use of the Newton-Raphson (or simply Newton) method, and is one of the 'best' rearrangements, being rather better than more obvious ones such as $x = 2/x$.

The equation given above has one very attractive feature. It is this. Suppose we estimate the value of $\sqrt{2}$ (it can be a very bad guess). Let's say 1.5. If we evaluate the right-hand side of the equation, using the value $x = 1.5$

$$\frac{1}{2}\left(1.5 + \frac{2}{1.5}\right)$$

we get 1.416666 . . . This is not equal to the left-hand side, so we know that 1.5 is not the square root of 2, but the new value is a better approximation to $\sqrt{2}$ than our first guess. This is always the case. Our guesses get better and better, until they are equal to one another up to the accuracy of the computer. We are executing an *iteration* based on the rearrangement given above. Formally, we represent this as

$$x_{n+1} = \frac{1}{2}\left(x_n + \frac{2}{x_n}\right)$$

x_1 is used as a guess to give us x_2, which is used in turn to give x_3 and so on. Using $x_1 = 1.5$, we get the sequence

$x_1 = 1.5$
$x_2 = 1.4166667$
$x_3 = 1.4142157$
$x_4 = 1.4142136$
and $x_5 = x_4$

As two values have coincided, clearly the sequence produces only this value, no matter how many times we repeat the iteration. Because of the problems of floating-point equality referred to in Chapter 1 (note p. 7), we usually set up some sort of test for 'equality within a tolerance' to tell us when the iteration is done.

Not all rearrangements of the basic equation provide this desirable property. Some converge on other values, some oscillate between two values without getting closer to the required value, some actually give poorer and poorer estimates, even when we start close to the answer. The branch of mathematics known as *numerical analysis* can be used to design suitable rearrangements with this property, and to analyze when and if they work well. It is a valuable technique, and you will see it in use in Section 5.3.

Power series

If you draw the graph of (say) sine x, and on the same axes draw the graph of

$$x - \frac{x^3}{3!} + \frac{x^5}{5!} - \frac{x^7}{7!}$$

you will discover that in a region close to the origin, the two curves are so similar that they appear to lie one on the other. As we move away from the origin, the 7th degree polynomial gets to be a poorer and poorer match for the sine curve, and in fact the two curves are only exactly on one another at the origin itself.

Again, you are not expected to be able to design such polynomials. This one is the *Taylor series* for sine x, and the technique used for deriving it can be applied to all functions, with the same mixture of success that we noted in the case of iteration. Sometimes a Taylor series gives a good approximation to the curve, sometimes it is better over a particular area, sometimes it doesn't work over bits of the graph—sometimes it works almost nowhere! (The method used here actually ensures that the Taylor series is a *perfect* approximation at one point—the origin.) A series such as this one is called a *power series*, because it is a series of ascending powers of x. Clearly it can be extended for as many terms as we choose (the pattern is easy to spot). The more terms we use in the calculation, the more accurate the result is as an approximation to the sine function. We work, therefore, with a convenient power series rather than attempting to evaluate functions directly from a definition.

Power series are convenient mathematically, but are also convenient computationally, because a quick and efficient method of evaluating them is available, called *nested multiplication*. This is discussed below, in Section 5.2; you will find power series used in the evaluation of functions in several of the following sections.

5.2 Nested multiplication

To evaluate the power series given for sine in Section 5.1, we should probably evaluate the coefficients 1/3!, 1/5! and so on, and use a high-level language statement such as

LET S = X − 0.166667∗X↑3 + 0.00833∗X↑5 − 0.0001984∗X↑7

This uses 3 exponentiations (raising to a power), 3 multiplications and 3 additions (or subtractions, which are essentially the same thing). For a one-off evaluation, this is quite satisfactory: it actually takes considerably longer to exponentiate than it does to multiply or to add (you will see why in Section 5.7), but in practice this does not usually matter much.

However, in the 'early days' when computers were very much slower than at present, there was much practical value in taking as many short-cuts as possible, and an efficient method of evaluating expressions such as the above, called *nested multiplication*, was frequently used. It uses only the operations of multiplication and addition, in the following way. To evaluate

$$a_n.x^n + a_{n-1}.x^{n-1} + \ldots \quad \ldots + a_2.x^2 + a_1.x + a_0$$

we would instead evaluate the equivalent expression

$$((\ldots(((a_n.x + a_{n-1}).x + a_{n-2}).x + a_{n-3}).x + \ldots \quad \ldots + a_2).x + a_1).x + a_0$$

Notice that we begin with the coefficient of the highest power. The *a*'s can be positive or negative, and some (as in the case of the sine series) may be zero. To evaluate the sine series, we would compute

((((((−0.0001984∗X + 0)∗X + 0.00833)∗X + 0)∗X
−0.16667)∗X + 0)∗X + 1)∗X + 0

Of course, we would not be so bound by ritual that we would actually use this expression, with all its redundant ' +0' expressions. We would of course write

```
LET Z = X*X
LET S = ((( −0.0001984*Z + 0.00833) *Z −0.16667) *Z +1) *X
```

This uses 5 multiplications, 3 additions and no exponentiations.

Nested multiplication is usually used in evaluating power series and polynomial expressions where they appear in the algorithms for intrinsic functions. As we have noted already, these are usually programmed (in a low-level language) to shave off every possible milli- or micro-second from the execution time.

Nested multiplication is enjoying a new popularity with the increased use of pocket calculators. As it uses only X, rather than powers of X, the use of a memory to store the current value of X for repeated recall can dramatically cut down on the number of key-strokes, and hence the chance of error.

5.3 The square root function

This algorithm uses both techniques from Section 5.1, and is in some ways a classic example of impenetrability. For example, it does some initial juggling to ensure that when a power series is used, the range of the X value is limited, so that a very good approximation is obtained.

The algorithm is usually called into operation by the high-level language statement

LET Y = SQR(X)

and it assumes that X is represented in normalized floating-point form. X can thus be written as

MANTISSA × 2^EXPONENT

If the exponent were even, say 2K, then the square root of X would be

SQR(MANTISSA) × 2^K

That is, we halve the exponent (an efficient one place right-shift, see Chapter 3) and take the square root of the mantissa.

Because the value is normalized, we know that the mantissa lies in a particular range: from $\frac{1}{2}$ to 1 (it cannot be negative, as we cannot take the square root of a negative number). We can thus choose the best power series approximation over this range. In fact, the best results are obtained by taking a linear approximation over this range—a power series with only one term, and using this approximation as the first guess in an iteration. Because the guess is known to be a good one, we can iterate a fixed number of times, rather than testing after each iteration to see if the accuracy is sufficient.

The algorithm is given in Fig. 37, and is discussed in detail beside it.

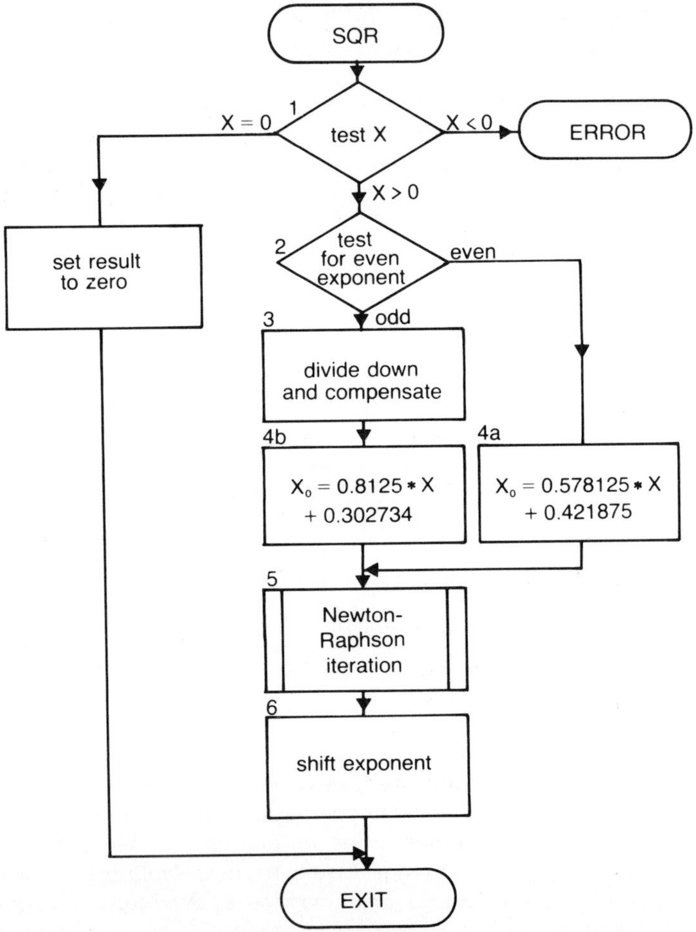

Fig. 37 A square root function

A square root algorithm

Step 1: Test for positive
If the input parameter (X) is negative, we cannot take the square root. If it is zero, we can avoid the intermediate calculation by exiting immediately with the result set to zero; otherwise, we proceed to Step 2.

Step 2: Test for even exponent
If the exponent is already even, there is nothing to be done, except to proceed to Step 4a. If it is odd, however, we correct this in Step 3.

Step 3: Divide down and compensate
With an odd exponent, we adjust the number by dividing by two, and compensating by increasing the exponent by 1. This unnormalizes the fractional mantissa, leaving it in the range $\frac{1}{4}$ to $\frac{1}{2}$.

Step 4: Obtain initial iterate
At Step 4a, the mantissa of the input parameter is between $\frac{1}{2}$ and 1; at Step 4b, between $\frac{1}{4}$ and $\frac{1}{2}$. In either case, we use the best available linear approximation over that range, to give us a first iterate for the subsequent iteration process.

Step 5: Newton-Raphson iteration
With a good first guess, our iteration (already given as an example in Section 5.1) converges rapidly. To save time-consuming checks on convergence, the iteration may be carried out a fixed number of times, possibly as few as twice.

Step 6: Shift exponent
From Step 2, and possibly Step 3, we know that the exponent is even. It may be halved by a one-place arithmetic shift to the right.

5.4 The trigonometric functions—SIN and COS

These two functions are intimately connected, in that COS(X) = SIN($\pi/2$ +X); accordingly, the cosine routine is simply an additional entry point into the sine routine. We simply add $\pi/2$ and then use the sine routine. This is shown in Fig. 39.

It is assumed that the input parameter X is a normalized floating-point value, representing the angle in radians. It is a simple matter to convert between degrees and radians.

Recall from Chapter 1, that the use of floating-point representation usually implies a lack of precision. For example, when trying to represent a number, it may not be possible to exactly represent the value, and a nearby approximation must be used. We also noticed that with very large numbers, the nearest approximation could be very far away (see Section 1.5). When precision becomes very low, the sine function (which repeats every 2π radians) cannot return a meaningful answer. In fact, as soon as precision is down to the point where adjacent representations are 2π apart, all angles will be represented by the same floating-point value, and the sine no longer distinguishes between them. The point at which this happens depends to some extent on the word-length in use, but in our example used in Chapter 1 (with 2 ×8-bit words), the value is as low as 800 radians. Well below this figure, the sine function is unacceptable in accuracy, so the algorithm includes a test and an error exit.

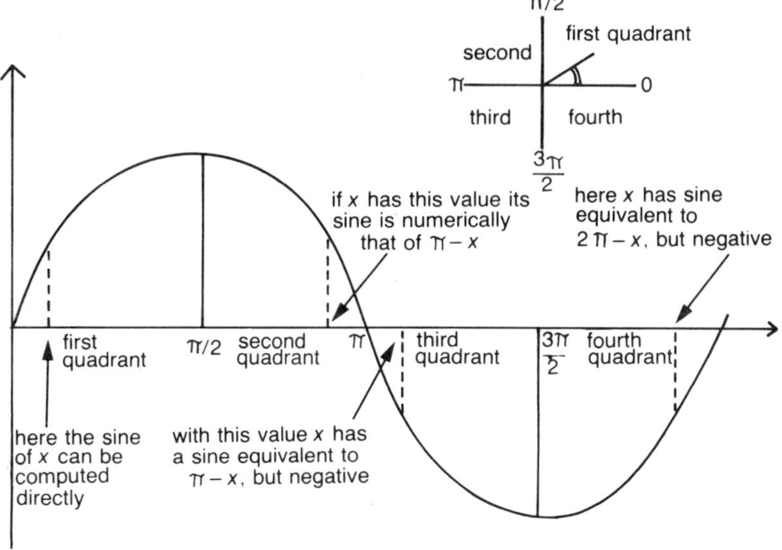

Fig. 38 The relation between sines of various angles

To avoid unnecessary work, a test is also included to see if X is very small (in absolute value); if this is the case, we may as well take sine X as being equal to X.

The method is to reduce the value of X to that of an equivalent angle in the first quadrant (between 0 and $\pi/2$). We can use a power series to give a good approximation to sine in this range, and then attach the appropriate + or − sign to the numerical answer. The algorithm is given in Fig. 39, and discussed below. Fig. 38 illustrates the mathematical steps involved.

A sine/cosine algorithm

Step 1: Test size of X
First we take the absolute value of X in order to test its size. If X was negative we remember this to adjust the later steps where appropriate). If X is 'large' take the error exit; if X is 'small' set the value to be equal to X (with appropriate sign) and exit.

Step 2: Reduce to first quadrant
Using the absolute value of X, repeatedly subtract 2π until the value is in the range $0-2\pi$. If it is now less than $\pi/2$ (in the first quadrant), it is this value of X that we pass to Step 3.
If X is in the second quadrant (between $\pi/2$ and π), then the angle π −X in the first quadrant has the same sine, and we pass this value to Step 3.
If X is in the third quadrant, its sine will be that of the equivalent angle π −X in the first quadrant, but the value

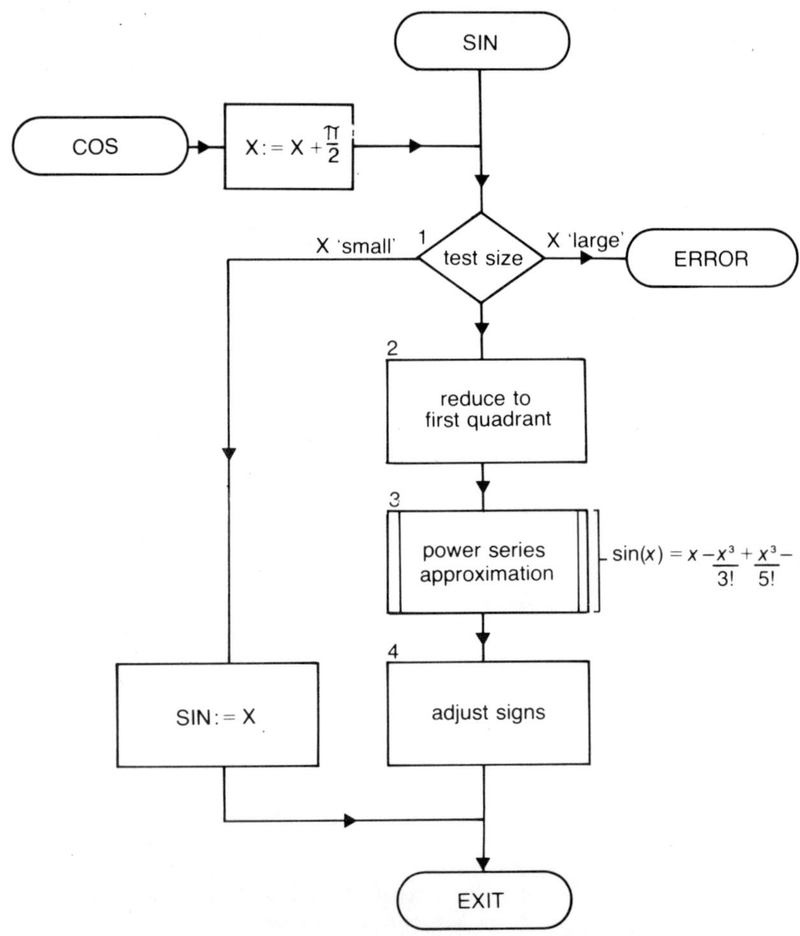

Fig. 39 A sine/cosine algorithm

will be negative. We have to 'remember' another minus sign for later adjustment.
If X is in the fourth quadrant, the equivalent angle is $2\pi - X$ and the value is again negative.

Step 3: Power series approximation
We now use a suitable power series approximation on the equivalent value of X. The coefficients—1/3!, 1/5! and so on will have been previously computed to maximum accuracy, and the evaluation will probably be by nested multiplication (see Section 5.2).

Step 4: Adjust signs and exit
Our current answer is a normalized floating-point representation that is certain to be positive. It may be necessary to negate the mantissa to compensate for any minus signs that have been neglected on the way.

5.5 The logarithm functions

In scientific computing, we require the values of two sorts of logarithms, those to base 10 (common logarithms), and those to base *e* (Naperian or natural logarithms). There is a simple relation between these, in that $\log_{10}(x) = 0.4342944822.\log_e(x)$. The constant is simply $\log_{10}(e)$ and is a consequence of the formula for changing the bases of logarithms:

$$\log_a(x) = \frac{\log_b(x)}{\log_b(a)}$$

Accordingly, the function for $\log_{10}$ is simply an alternative exit point from the $\log_e$ algorithm, and often is not provided at all. The method is similar in spirit to that used for the square root function: we juggle with the exponent of the floating-point representation, in order to be able to use a power series approximation on the restricted range of the mantissa.

As before, we assume that the argument X is a normalized floating-point representation, and can be written as

$$X = \text{MANTISSA} \times 2^{\text{EXPONENT}}$$

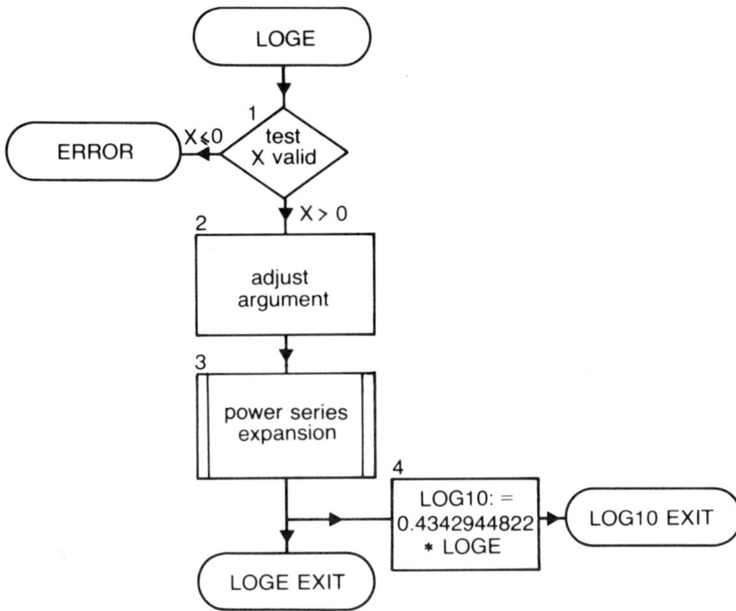

Fig. 40 A logarithm algorithm

By the usual rules for logarithms, we then have

$$\begin{aligned}\log_e(X) &= \log_e(\text{MANTISSA}) + \log_e(2^{\text{EXPONENT}})\\ &= \log_e(\text{MANTISSA}) + \text{EXPONENT}.\log_e(2)\\ &= \log_e(\text{MANTISSA}) + 0.6932.\text{EXPONENT}\end{aligned}$$

Effectively, this takes care of the exponent, and leaves us to evaluate the logarithm of the mantissa; as the mantissa is 'small', it will give a good approximation with the power series that we shall use. There is just one problem. The exponent is stored as a fixed-point integer, and the constant needs to be in floating-point form. We do not presently know how to multiply these together. This will be covered in Section 5.8. Unfortunately, the series is not as simple as that for the sine function. There is no Taylor series expansion for $\log_e(x)$; for technical reasons, we obtain a series for $\log_e(1+x)$. But even this is unsatisfactory, because it has only a very limited range over which it gives the correct answer. However, a series that does work for all values of x is

$$\log_e(x) = 2.\left(\frac{x-1}{x+1}\right) + 0.66667.\left(\frac{x-1}{x+1}\right)^3 + 0.4.\left(\frac{x-1}{x+1}\right)^5 + \ldots$$

The coefficients follow the pattern 2, 2/3, 2/5, ... 2/(2n −1). These will be turned into their decimal equivalents, as above, to avoid unnecessary divisions. The algorithm for the common and natural logarithm functions is shown in Fig. 40; we discuss it here in detail.

A logarithm algorithm

Step 1: Test for valid argument
X cannot be negative, or zero. If it is, we take the error exit. (If X is 1, we may by-pass the algorithm, setting the result to 0 directly.)

Step 2: Adjust argument
In order to simply use nested multiplication on the power series, we compute the intermediate value

$$Y = \frac{(\text{MANTISSA} - 1)}{(\text{MANTISSA} + 1)}$$

Step 3: Power series expansion
$\log_e(x)$ is now evaluated as
LOGE =2∗Y +0.6667∗Y³ +0.4∗Y⁵ +EXPONENT∗0.6932

Step 4: Common logarithm exit
If necessary, we multiply the result by 0.4342944822 to get the common (base 10) logarithm, and exit.

5.6 The exponential function

This algorithm is a classic example of deviousness. There is a very well-known Taylor series expansion for the function exp(x) or e^x. It is, indeed, one of the simplest of all Taylor series, and it is valid for all values of x:

$$exp(x) = 1 + x + x^2/2! + x^3/3! + x^4/4! \text{ and so on} \ldots$$

Unfortunately, with large values of x, it requires many more terms of the series to produce an acceptable answer than it does with small x. The algorithm, accordingly, goes in for some quite terrifying manipulation in order to be efficient.

It is very important that the reader realizes that it is worthwhile for a manufacturer or designer of software to devote effort to developing algorithms of this complexity. The resulting software will be used many, many times over. In a user program, it is important to remember that the programmer's time is as scarce and costly a resource as computer power—indeed more so!

The reader may well ask why the exponential function is not simply evaluated as e^x (i.e. 2.7182818 . . . to the power x). The answer is that the algorithm for raising numbers to a power uses this exponential algorithm!

The number e, and in particular the function e^x or $\exp(x)$ is important to mathematicians, engineers and scientists partly because all power functions, such as 2^x, 3^x and so on can be written as e^{kx} for some suitable choice of k. For example, $2^x = e^{0.6932x}$, $3^x = e^{1.0986x}$. On the contrary, e^x can always be written as 2^{kx}, for some other suitable k. In fact, $e^x = 2^{1.4427x}$.

The first step in our algorithm is to look at $2^{1.4427x}$. This can be written as

$$2^{(\text{INTEGER-PART} + \text{FRACTIONAL-PART})}$$

That is, separating the whole number part from the part following the decimal point. For example, if x was simply 1, we would have $2^{1 + 0.4427}$. By using the expression given below for evaluating $2^{\text{FRACTIONAL-PART}}$ taking the floating-point result and adding INTEGER-PART to the exponent, we can directly evaluate e^x. As we hinted before, the range of the FRACTIONAL-PART is limited, so this unlikely-looking expression gives a very good approximation.

$$2^F = 1 + \frac{2}{(9.95460 + 0.03466 * F^2 - F - (617.97227/(F^2 + 87.41750))}$$

The algorithm has been described in some detail above, but for completeness, we show it in flowchart form in Fig. 41, and summarize it below.

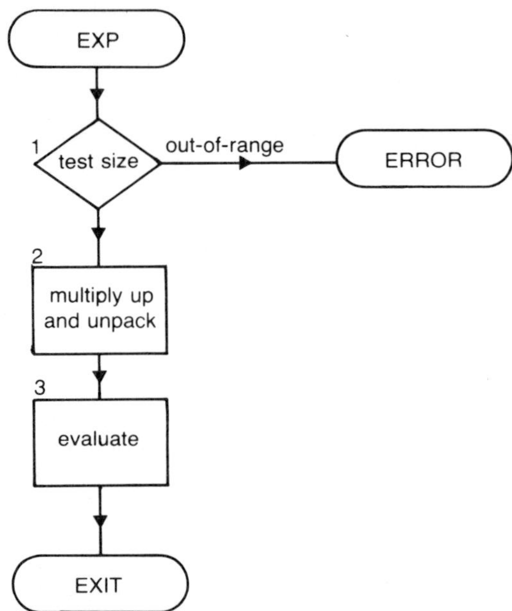

Fig. 41 An exponential algorithm

An exponential algorithm, exp (x)

Step 1: Test size
With a large value of X, floating-point exponent overflow will result; if X is large and negative—underflow. At this step we test for these conditions (they depend on the word-length) and exit if necessary.

Step 2: Multiply up and unpack
Multiply X by 1.4427 . . . and separate the integer and fractional parts, using the techniques of Section 5.9.

Step 3: Evaluate
Use the expression above to evaluate $2^{\text{FRACTIONAL-PART}}$ add INTEGER-PART to the exponent and exit.

5.7 Power functions

These are an exception to the general rules for intrinsic functions. In order to raise *a* to the power *b*, we do not need to call a function, in the manner of SQR or SIN. Instead, we usually write

LET X = A↑B
or X = A**B

In order to evaluate an expression such as this, we make use of both exponential and logarithm algorithms. Mathematically speaking:

$$A^B = \exp(B \times \log(A))$$

Hence to perform an exponentiation, we take the $\log_e$ of the base, multiply by the power and take 'antilogs' (that is, evaluate the function exp).

5.8 Mixed-mode arithmetic

In a computer that uses both fixed-point and floating-point storage conventions, some numbers—the smaller integers—can be represented as either a floating-point or a fixed-point quantity. For example, with the conventions of Chapter 1

00010110 and 0 1011000 01111110

both represent +22 (remember that the exponent is in excess-128 storage).

It is sometimes necessary to convert between these representations: for example, see page 47. In order to operate on two numbers that are stored in different ways, we must convert one of them to the same system as the other. In multiplying an integer by a fraction, we must represent both in floating-point form.

FORTRAN is an example of a language where this choice of methods of representation is made explicit: we shall discuss this below.

The FLOAT function

This takes a fixed-point integer and turns it into a floating-point quantity. We simply construct a floating-point quantity with +7 exponent (in our 8-bit example) and the integer as its mantissa. In general, the result will not be normalized, so we apply the normalization algorithm of Section 4.4.

The FIX function

This takes a floating-point quantity, and returns a fixed-point representation. Where the original value was an integer that is small enough, the result is the fixed-point representation that is numerically equal to the original value. If the value was too large, garbage will result. If the value was not an integer, but was still within the integer range, the algorithm achieves a similar effect to INT, discussed below.

The algorithm takes the original value and shifts it, unnormalizing it in the process, until the exponent is +7. The mantissa can then be treated as a fixed-point quantity. It may be desirable to provide an error check and alternative exit to detect values too large to satisfactorily process.

A note on languages

Many languages allow the programmer to explicitly choose which method of storage is to be used for numeric quantities. FORTRAN is a good example. All variable names beginning with a letter in the range I to N (e.g. NUMBER, ICOST, JTOT, K) and all those names explicitly referred to in the INTEGER statement, e.g.

INTEGER COST, TIME, X

are said to be of type INTEGER, and fixed-point storage is used for the quantities stored therein. An attempt to assign a non-integer such as 2.5 to an INTEGER location, as in

NUMBER = 2.5

is not illegal, but is rather unexpectedly equivalent to FIXing the value 2.5 (giving 2) before the assignment.

INTEGER constants are written without a decimal point: 2, 32767, 1024, etc. An arithmetic operation on two integers gives an answer that is effectively FIXed before it is made available. Thus

NUMBER = 25/4

assigns the value 6 to NUMBER, rather than 6.25

All other variable names, and those declared in the REAL statement, are of type REAL and use floating-point storage. As constants, they are written with explicit decimal points. Thus 3 is an INTEGER, while 3.0 is REAL.

In early standard FORTRAN, mixed-mode arithmetic, such as

25/4.0

was not allowed, although unambiguous. Many compilers actually did allow this, under the principle that what was unambiguous should be permitted. The result would be the REAL number 6.25, although assigning it to the location NUMBER would truncate it to 6, as discussed above.

BASIC is, in fact, one of the few languages that does not always allow this explicit control, although later compilers and interpreters have been introducing an integer-variable convention, usually by the addition of a '%' to the variable name, or to the constant. Thus, A%, Z9%, 1%, 5% would all be integers. In a small computer, the saving of space that results can be exceptionally valuable.

5.9 Unpacking numbers—INT, CHOP and FRAC

Although the FIX function of Section 5.8 can be useful, in that a side-effect of it is to extract the integer part of a mixed number (e.g. FIX(3.5) = 3), this can be done more usefully with a separate routine that produces a floating-point result. In any case, there is a need for a routine such as this in BASIC where there may not be an integer-type variable.

There is a certain ambiguity about the 'integer-part function' when applied to negative numbers. The question is whether FIX(−3.5) is to be −3 or −4. It is mathematically most convenient to define a function INT(X) as the largest integer not greater than X; in this case INT(−3.5) is −4.

However, reference is sometimes made to a hypothetical integer function CHOP(X), where CHOP(−3.5) is −3. It is not difficult to simulate this function using INT and the absolute value function ABS(X) in the following way:

CHOP(X) = INT(X) if $X \geq 0$
= −INT(ABS(X)) otherwise

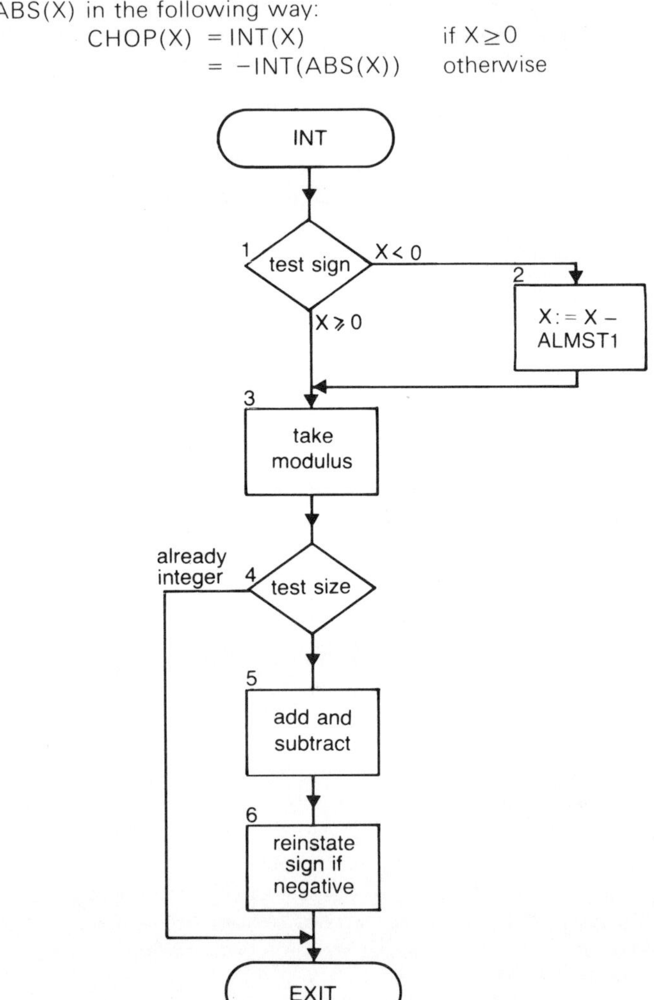

Fig. 42 An integer-part (truncation) algorithm

Given a function to extract the integer part of a number, it is a simple matter to obtain the fractional part, in the following way:

FRAC (X) = X − CHOP(X)
or = X − INT(X)

depending on how one wishes to define FRAC(X). Because of the ease of writing functions to obtain CHOP(X) and FRAC(X), it is customary to provide only INT(X) as an intrinsic function.

The algorithm given in Fig. 42 is delightful in its simplicity. Basically, it consists of adding a very large number, and then subtracting that number again; the number (depending on word size) is chosen so that at the normalization stage that takes place after the addition, all the fractional part, but none of the integer part, is lost. There is a slight complication to do with the definition of INT(X) for negative X.

A floating-point truncation algorithm

Step 1: Branch on sign
It is necessary to treat positive and negative values separately for the INT function; this would be unnecessary for a CHOP function, which could work with the absolute value.

Step 2: Adjust for negative argument
In order to obtain the correct answer for a negative argument, the constant ALMST1 is subtracted. This value is the closest representation below 1 that can be stored in the computer. It depends on the word size.

Step 3: Take modulus
We work hereon with the absolute value. If necessary, the minus sign is replaced in Step 6.

Step 4: Test size
Beyond a certain size of exponent, normalization will have already removed any fractional part, and the value is already integral. For an 8-bit mantissa, the exponent should be less than 8.

Step 5: Add and subtract
This is the core of the algorithm. We add, normalize (as part of the addition) and subtract. With an 8-bit mantissa, there are seven binary places of precision. Adding 2^7 loses all the fractional information.

Step 6: Negative?
If a minus sign was removed at Step 3, it is reinstated here.

5.10 Some other functions

The functions of this last section are either so simple to program in a high-level language that they are often not provided as intrinsic functions, or are so fundamental that they are implemented by whichever method is convenient. For example, in the DECSystem-10 computer, the ABS function is a side effect of the basic MOVE instruction; another computer would use a test and multiplication and so on.

The absolute value, or modulus function ABS

This simply ignores the sign of the number. At worst it should never be more complicated than a test for positive/negative followed by multiplication by −1.

The SGN function

This just returns +1 for a positive argument, 0 for a zero argument and −1 for a negative argument. As a sequence of high-level statements, it can be a test for zero, followed by

sgn(x): = **if** 0 **then** 0 **else** abs(x)/x;

The factorial function

This function is something of a classic example, referred to in many text books to illustrate the difference between an iterative and a recursive approach to programming. The function FACTORIAL(X), usually written mathematically as $x!$, is defined for non-negative integers and is simply the product of all positive integers less than or equal to x, that is

$$x! = x.(x-1).(x-2) \ldots \quad \ldots 4.3.2.1$$

Thus 3! = 3.2.1 = 6, and 5! = 5.4.3.2.1 = 120. By convention, 0! = 1.

We may compute the factorial of an integer by means of an iteration in the following way:

```
100 REM: THIS ROUTINE PLACES X! IN F
110 LET F = 1
120 FOR I = 1 TO X
130   LET F = F*I
140 NEXT I
```

In a language which permits recursion (see *System Software* by P. J. Barker) the work can be done by a statement such as

factorial (x): = **if** $x = 1$ **then** 1 **else** $x*$factorial(x =1);

Both these approaches are pointless! In our 8-bit computer, used for examples in the early chapters, 6! is already greater than the range permitted for fixed-point integer storage, and 12! is too large for 8-digit decimal storage (a typical pocket calculator range). It is more efficient (certainly in time, possibly in space) to provide a small look-up table for the few exactly representable factorials, and to treat larger values either as errors, or to use an approximation known as Stirling's formula:

$$x! \simeq \sqrt{(2\pi n)}\left(\frac{n}{e}\right)^n$$

where e is the base of 'natural' logarithms, 2.71828 ... This approximation gives quite reasonable order-of-magnitude answers (see Fig. 43).

x	factorial (x)	
	By multiplication	By Stirling
12	4.79×10^8	4.76×10^8
20	2.43×10^{18}	2.42×10^{18}
30	2.65×10^{32}	2.65×10^{32}

Fig. 43 Comparison of factorial evaluation

This function is almost never provided as an intrinsic function, and is usually left to the programmer. A way of avoiding it is mentioned in Section 7.4.

Quotients and remainders

It is not at once clear what results are required for the quotient and remainder on dividing X by Y, where one or both are negative. The high-level statements for positive X and Y are

```
LET Q = INT(X/Y)
LET R = X - Q
```

Modular arithmetic

As the evaluation of an expression *modulo n* is a matter of obtaining the remainder on division by n, this can be handled by the statements given in the last paragraph.

Divisibility tests

To test if the integer Y divides into the integer X, it is sufficient to use the statements of the last two paragraphs to determine if the remainder on division is 0.

Summary

This chapter has collected a number of results indicating how some commonplace arithmetic functions may be provided in a compiler library. The list of functions covered is similar to that found in any compiler, except that the functions TAN, ARCTAN, COT, SEC, COSEC, etc., have been omitted as being less relevant than those included. The functions that may be found in the preceding sections are listed in the index under *Functions*.

EXERCISES

Public examination question

1. *Three programmers are discussing the provision of a routine to compute the value of* n!
   *(*n! *is factorial* n*, defined, for any positive integer* n*, as the product* n × *(*n − *1)* × *(*n − *2)* × . . . × *2* × *1.)*
   Their conversation is as follows:
   Programmer A: *'I propose that we write an iterative program.'*
   Programmer B: *'That seems wasteful:* n! *can easily be defined recursively, so I propose that we write a recursive program.'*
   Programmer C: *'Nonsense. 12! is already outside the range of fixed-point integers representable in this computer. I propose a table look-up approach.'*

Write three paragraphs to explain to someone, who did not know the technical meaning of the underlined words, the essential differences between the three approaches to the problem.
(London, 1979) *(15%, 3 hours)*

6 Random Numbers

There is no such thing as a random number. And yet, the significant omission from the list of standard functions in the last chapter was the one invoked in a statement such as

100 LET X = RND(Y)

(For syntax reasons, some implementations of BASIC, in which this example is given, require that RND be given an argument even though Y is purely a dummy variable and is not used at all.)

What does this function achieve, and how does it do it? As generally implemented, it assigns to X a number in the range 0 to 1. Chosen in this way, the number satisfies most people's understanding of the word random; so what does it mean to say that there is no such thing as a random number?

The problem arises over our carelessness with language. Randomness is not a property of numbers, but of *sequences* of numbers. Thus the result of throwing a die could be any of the integers 1,2,3, . . ., 6. It would not surprise us if it were a 5. But if we threw a sequence of dice throws, the sequence

1,6,3,5,4,3,4,2,3,6,4,2,1,5

would pass unremarked, while the sequence

5,5,5,5,5,5,5,5,5,5,5,5,5,5

would indicate something peculiar about the die.

We should say that the first sequence is a *random* sequence, while the second is not. That is to say, we would be prepared to accept that the first sequence was likely to have arisen 'by chance', if the only forces affecting the die were those found by experience to affect fair dice. Fourteen throws of a fair die would be very unlikely to give us fourteen 5's. We have reason to believe that the die is not fairly balanced, is not being recorded accurately, or is being manipulated in some way.

But there is another problem. While our first sequence is 'random', we should be rather perturbed if the next batch of 14 throws produced the same sequence. In fact, both 14-digit sequences are equally unlikely. There is just about 1 chance in 100 000 000 000 that either will arise in 14 tosses of a die. So 'randomness' is not a description of the likelihood of a sequence occurring. Rather, it is a measure of the 'reasonableness' of the digits in the sequence. The first sequence is reasonable, because (roughly) it has about equal numbers of each of the six digits. The second is not, because fourteen 5's with no 1's, 2's and so on, does not strike us as reasonable.

This chapter is firstly about how to test a sequence of numbers for randomness (or reasonableness), and secondly, about how to generate a sequence of numbers that is 'random'. We shall admit immediately that we shall not be able to generate truly random sequences, and shall be forced to compromise.

6.1 Testing for randomness

Let us begin with our original pair of sequences. The first appears to be random, because it is reasonable to expect that, in the long run, equal or nearly equal numbers of each of the six digits will appear. This gives us a first test for randomness.

Testing for single-digit frequency

In a sequence of N random integers, in the range 1–m, we expect the proportion of 1's, 2's and so on to be approximately N/m. Thus 100 dice throws should yield about 16 or 17 of each digit. We shall not be surprised to find minor variations from this pattern, and a statistician would actually tell us what qualified as a 'minor' variation. But whatever characteristics it might otherwise possess, any sequence that we claimed was random would have to satisfy this test.

Let us look at another sequence of dice throws:

1,2,3,4,5,6,1,2,3,4,5,6,1,2, . . .

This satisfies the single-digit frequency test, but its repetitive, ordered nature suggests a lack of randomness. A further test is appropriate.

Testing for digit-pair frequency

In our latest sequence, we might characterize the lack of randomness by observing that the digit 1 is always followed by a 2, a 2 by a 3 and so forth. Whereas, in a random sequence, we should expect about equal numbers of digit pairs. That is, we look at the first and second digit together and note it (12). Then the second and third (23) and so on. In a random sequence of N integers (in the range 1–m), there are N-1 digit pairs. We should expect these N-1 to be distributed evenly among the m^2 possible values. That is, in the long run, we should expect to find about $(N-1)/m^2$ of each digit pair.

Clearly this method of testing can be extended to triplets, quadruplets and so on. A random sequence must satisfy each test. It is easy to construct a sequence satisfying the test for (say) digit pairs, that fails the triplet test:

1112131415162122232425 26 . . .

This observation is common to all tests for randomness. No one test is sufficient. The whole battery of tests given here must be applied to a possible random sequence, and it must pass them all, and any others that you can think of as well.

Testing for gaps

After a 5 in our sequence of dice throws, we should neither expect another 5, nor be surprised to see one. But in the long run we should be very surprised to find one 5 following another regularly, or indeed to find that we had to wait for 100 throws before another 5. We could monitor the gaps between successive 1's, 2's and so forth. We would expect some gaps of 0 (that is, successive digits the same) and a very few gaps of 100 or more. But by and large, we should expect the average gap to be around 6. Again, the statistician could tell us what deviation from this average would be permissible.

The coupon-collector test

If our individual digits were not emerging from some random process such as the BASIC RND function, but were being given away with gallons of petrol, and we were collecting them, then we should be surprised (and probably pleased) to have accumulated a complete set of 6 after only 6 digits. Equally, waiting for 100 digits before completing a set would be unusual (and frustrating). We can monitor the length of time that it takes to accumulate a set, starting at the first digit in the sequence, then the second, and so on. There will be few 6's and fewer 100's. The statistician will tell us what is reasonable.

Testing for runs

This is yet another investigation that we should expect to have confirmed in a useful random number generator. Starting at the beginning of the sequence we count how many digits increase. The lengths of such increasing sequences, starting at each digit in the sequence, should show a statistically acceptable distribution.

Some other tests

A test of some curiosity value is the *poker player test*. Here we look at adjacent quintuples of digits and count the occurrences of simple analogues of poker hands (five-of-a-kind, full-house, two pairs, and so on).

The *permutation test* argues that in any sub-sequence of 5 adjacent digits, 12345 should appear about as often as 53145, and any other sequence. It is essentially a generalization of the first two tests.

The perceptive reader should detect two distinct styles of test in the above. Those that examine the expected numbers of certain digit combinations (digit-pairs, poker hands, etc.); and those that look at the distribution of certain patterns (run tests, gap tests, etc.). It would be a useful exercise, extremely instructive, and of some value in practice, to devise a test of a totally different nature.

6.2 Computer-generated random numbers

If we believe that the preceding discussion has told us how to recognize a reasonable string of random numbers, we are still left with two questions: what do we want them for, and how are we going to get them?

There are many applications of random numbers. Some capitalize on the unpredictability of the next digit while others make subtle use of some of the other properties we have mentioned. Lotteries and sweepstakes require that the number, ticket, or name to be chosen is picked in a manner that precludes any foreknowledge of what it is likely to be. There is little improvement on the traditional 'names in a hat' method, where the number of items to be chosen from is suitably small. For major lotteries, this method is unworkable (few hats are large enough) and probably not secure enough against sleight of hand. The British Premium Bond weekly lottery uses random emissions from a stimulated gas to generate truly random numbers, and certainly no fault has been found with the performance of this method in over 20 years of regular service.

From our point of view, a more important use of random sequences of numbers is in a variety of simulations of physical and social systems. The name 'Monte Carlo method' is a whimsical description of a serious aspect of statistical verification of otherwise unobtainable results, in which decisions are repeated a large number of times, with the outcome being determined by a sequence of random numbers. Thus, in simulating the diffusion of liquids, we might choose to look at the molecules of the liquids involved. Some probabilities for the possible movements and transfers are then assumed, and then molecules chosen at random according to these probabilities. Capitalizing on the computer's ability for fast repetitive computation, a large number of microscopic changes can be simulated, and the results on the macroscopic level compared against scientific observation of the 'real world'.

The traditional methods of generating random numbers—names in a hat, toss of a coin, roll of a die, are all suited to the generation of small sequences of random numbers. Experience shows us that with well mixed hats, fair coins and unbiased dice, such methods provide sequences that exhibit the properties we have catalogued as being desirable in random sequences. However, we cannot in practice call on our computers to toss coins or to reach into hats; accordingly, much effort has been expended over the design of programs to provide random number sequences.

Pseudo-random numbers

Immediately, we have a problem. Every single component of a digital computer is absolutely and completely determinate. Whatever piece of random-sequence generating software we write, every single time that it is executed, it will produce exactly the same sequence of random numbers. This would appear to contradict our definition of randomness. We suggested at the very beginning that if our random sequence was always the same, we should cast doubt on the fairness of the method used to generate it.

We must not confuse randomness with complexity. If we design an algorithm that starts with an 'unlikely' number, like 2.0121945, multiply it by something like 0.16111944 (my computer thinks that this is 0.1061978, but it may not have read Chapter 7 yet!), take a square root and double the result, then our answer is in some way unpredictable, but unfortunately when we next call upon the algorithm, it will remorselessly produce the same result, again, and again.

We solve this problem obliquely. We first of all aim for a sequence of 'pseudo-random' numbers. That is, a sequence that satisfies the tests outlined earlier in this chapter, but which will be repeated on subsequent execution of the algorithm. We aim for a very long sequence, before any substantial repetition sets in. If our simulation, or whatever, was only to be performed once, then this sequence would, by itself, be sufficient. But we then aim for something else—we try to determine the starting point in this sequence using some feature of the computer system that is not predictable. We may look at the time-of-day clock, for example, and use its current value to mark a starting position in some way. Or we may share the random number routine among several users of a time-shared computer, and let each user in turn take the next available number in sequence. Or we may look at some particular location in store and use whatever is stored there left-over from some other processing, as if it was a number telling us where to start. Neither of these methods alone is a good generator of random numbers, but together with a well-designed pseudo-random sequence, acceptable results may be obtained.

Pseudo-randomness is not always a disadvantage. It can be helpful in debugging simulation programs, to repeat the same sequence of numbers and only 'turn on' the random-start feature when going for a full production run.

We find various methods of controlling this random-start. For example, some versions of BASIC provide a statement

```
100 RANDOMIZE
```

which may be omitted during test runs, to provide the same sequence repeatedly, but inserted when 'truly random' numbers are required. The same effect may sometimes be achieved by utilizing the dummy parameter insisted on in some languages, for syntax reasons. Thus, in some versions of BASIC

```
100 LET Y = RND (0)
```

provides a 'truly random' sequence, while

```
100 LET Y = RND (5)
```

starts the sequence at a fixed point determined in this case by the parameter 5. It is customary, in many implementations, for the statement

```
100 LET Y = RND (X)
```

for a negative X, to return not a random number, but the value of the starting-point, so that it may be saved, and the sequence restarted at the same point if so desired.

Some theoretical considerations

It is not normal to produce random integers from a random-number generator. Most practical algorithms produce fractions in the range 0 to 1. It is easy to convert these into the desired integers, and a technique is given in Section 6.4. There is a good reason for this—the methods commonly employed tend to produce numbers that are 'more random' at the left-hand (most-significant) end, than at the right. By treating the number produced as if it has a point at the left-hand end, this effect is minimized. It is not necessary to fully understand why this happens—one example of the type of problem that arises will provide an illustration.

If, for example, a step in the generating algorithm is the multiplication of two numbers, and one of them is even, then the product will be even. Subsequent multiplications cannot remove this evenness, and indeed, if at any stage we multiply two even numbers, the product will be divisible by four, as will all subsequent numbers produced in this way. So the effect of division properties such as evenness, tends to propagate, producing a non-random sequence. As we said above, treating the number generated as a fraction minimizes this.

Another consideration is that the sequence we produce will be *cyclic*: that is, after some period it will return to the start, and generate the same numbers over again. It cannot be otherwise! Practical random-number generators use the previous number as the start of the algorithm to generate the next. If at any stage the sequence returns to its original starting value, the sequence will repeat exactly from then on. The sequence must return, because of the fundamental limitation on numeric processing that we observed back in the very first chapter. With 8 bits/word, there are only $2^8 = 256$ different bit-patterns, and hence representable numbers. The 257th random number generated must be one of the previous 256, even if the sequence has not repeated up to that point.

6.3 Practical random-number generators

There are a number of methods for generating sequences of random numbers—more strictly pseudo-random numbers, but this is understood in what follows, and we shall discuss two of the most commonly used methods.

The middle-square (or centre-square) method

Given a particular member of the random sequence being generated, this method produces the next member in the following way.

1. square the existing term, producing a double-length result; (Although the existing term has been presented to, and used by, the programmer as a fraction, it is treated by the random-number generator as an integer for the purposes of generating the double-length square.)
2. extract the middle digits from this result, discarding the remainder (this can be achieved by masking and shifting).

The process can best be illustrated by an example using 4-digit denary numbers:

initially:	1234
square:	01<u>5227</u>56
extract:	5227
square:	27<u>3215</u>29
extract:	3215

Thus, this sequence produces the terms

0.1234, 0.5227, 0.3215, 0.3362, 0.3030, 0.1809, . . .

Fifty iterations later, this sequence runs into trouble. Having just produced 0.3317 . . .

square:	11002489
extract:	0024
square:	00000576
extract:	0005
square:	00000025
extract:	0000
square:	00000000
extract:	0000

and clearly the sequence now produces nothing except 0000.

Zero is an easy value to avoid; it may be detected when it occurs, and the sequence restarted from somewhere else. But it is hard to detect and correct a sequence that cycles too frequently to be of any use, such as

0.6100, 0.2100, 0.4100, 0.8100, 0.6100, . . .

This is not to say that the middle-square method is unworkable, but that it must be used with care.

The linear congruential method

The name of this method may appear frighteningly scientific, but it describes a method that is particularly convenient and efficient in practice, and which has the additional advantage of depending on four constants, which may be adjusted to tune the sequence to maximum randomness, and maximum length of sequence without repeat.

Given the existing term of the sequence, the next term is generated in the following way:

1. compute a new value by multiplying the existing value by the constant a and add a further constant c
2. divide this result by m (another constant), discard the quotient and present the remainder as the next random number in sequence.

As before, the algorithm treats the bit-patterns internally as integers, but presents them to the user as fractions.

This method may be expressed symbolically as

$$X_{i+1} = (a.X_i + c) \bmod m$$

where 'mod m' indicates the operation of taking the remainder after division by m (this is discussed in Chapter 5).

The constants a, c, and m, together with the starting value X_0 provide us with four values which may be adjusted to give a useful sequence. For example, let us again consider a denary example. With $X_0 = 1234$, $a = 57$, $c = 101$, and $m = 943$, we get the following:

initially:	1234	
×57	70338	
+101	70439	
mod 943	657	i.e. $X_1 = 0.6570$
×57	37449	
+101	37550	
mod 943	773	$X_2 = 0.7730$

There is nothing special about our choice of constants in the above; however, the choice of each has a significant effect on the sequence generated. Perhaps the easiest effect to see is that of the modulus constant m.

The only possible remainders after division by m are the numbers 0, 1, 2, 3, ... (m-2), (m-1). There are, of course, m of these, so that m effectively defines the maximum possible length of the random sequence. Whether the maximum possible length is attained or not will depend on the other constants.

As the maximum possible length of sequence imposed by the word-length of the computer is 2^8 (with 8 bits/word) or in general 2^N (with N bits/word), there is little point in choosing m larger than 2^N. It is also particularly convenient if we choose m actually equal to 2^N. In that case, the division by m is automatically effected by ignoring all bits in excess of the least significant N—a full word's length, in fact. Indeed, if we ignore all the bits that overflow into the higher-order word during fixed-point multiplication, we are effectively performing arithmetic modulo N.

Because of this convenience, m is usually taken to be 2^N. But it is only marginally more difficult, using masking operations, to work with any power of 2. A choice of m which is not a power of 2 will significantly increase the processing time, however, because of the need for a further fixed-point division, and the identification of the remainder.

The choice of a is dependent on the choice of m. Experience and theory show that (in a binary computer), a should lie between $\sqrt{m}$ and $m-\sqrt{m}$. (For details, consult: D. E. Knuth *The Art of Computer Programming* Vol. II: Seminumerical algorithms, Section 3.2.1.) Thus, in our earlier example, with $m = 943$, a should lie in the range from 30.708 . . . to 912.292. In the case of our 8-bit/word computer, with m chosen as 256, the range would be from 16 to 240. It is also mathematically convenient, and provides good random sequences if a is of the form $8k+5$ for some integer k. That is, $a = 5, 13, 21, 29, 37$, or other similar values. Thus a better choice of a than the 57 used in our example above, would be 53, or 61. The constant c should be chosen (in a binary computer) to be odd. It acts to counteract the propagation of evenness referred to earlier. Choosing c larger than m is pointless.

The final constant X_0 is of lesser importance, in that it does not affect the properties of the resulting sequence to such a great degree. To avoid having to store an extra value as part of the routine, it is common to set the value of X_0 to be the same as a. The first operation, the first time that the routine is called, will thus be to square the value of a.

6.4 Obtaining useful random sequences

With a well-chosen random-number generator, we can be confident that our sequence will be of maximum length, and will satisfy as many as possible of the tests outlined in the first section of this chapter. The question remains: how do we obtain random sequences in the range we require, and distributed as we wish, from the basic quantity provided—fractions in the range 0 to 1? It is not enough to simply use the RND function (say) and scale the value provided for us so that it lies in the correct range. If we are simulating the throw of a die, we require random integers where any of the 6 possible integers are equally likely. If we are simulating a queue in a supermarket, we shall need random integers weighted to make it more likely that 1 person enters the shop at a given moment of time than that 50 do!

Fortunately, it is relatively easy to write routines to meet the commonest needs. The mathematical and statistical justification of the algorithms given here, is best followed in a specialist textbook on statistics, or in one of the references in the bibliography.

Uniformly-distributed integers in a given range

Given a fraction in the range 0 to 1, we can obtain an integer in the range 1 to 6 by a statement such as the BASIC

```
100 LET X = INT(6*RND +1)
```

In general, to obtain integers uniformly distributed in the range A to B, we evaluate

```
100 LET X = INT((B -A +1)*RND +A)
```

It may of course be necessary to supply the RND function with an argument, as mentioned earlier.

Normally-distributed random numbers

If we are simulating some natural measurement, such as height or weight, we usually require random numbers (not necessarily integers) that have a peak about a central mean value, and which die away to either side of this value. Such a distribution is called a *Normal* or *Gaussian* distribution.

A Normal distribution is characterized by its *mean* (the value about which it is centred) and its *standard deviation* (a measure of its spread). Different values for these two quantities give different distributions of values, but all with the same basic shape.

The algorithm given below generates a standard form of the Normal distribution: the one whose mean is 0 and whose standard deviation is 1 (it thus includes as many negative numbers as positive ones, and the values cluster around zero). This distribution is a sort of parent distribution for all Normal distributions, because we can obtain values from another Normal distribution with mean M and standard deviation S by simply evaluating

```
100 LET X = S*N +M
```

where N is a Normal random number from the parent distribution.

The algorithm shown in Fig. 44 and discussed in detail below, actually produces a pair of standardized random Normal numbers. The second value can be stored and used the next time a value is required.

Step 1: generate uniformly and adjust

We take two uniform random numbers in the standard range 0 to 1, and by executing a statement such as

```
100 LET V = 2*RND -1
```

produce random numbers (not integers) in the range −1 to +1.

Step 2: test sum-of-squares
Square each of the two numbers and add them; if the sum, S, is greater than 1, reject the values and return to step 1 for another try.

Step 3: evaluate
The two expressions

$$V_1\sqrt{\frac{-2\log_e S}{S}} \text{ and } V_2\sqrt{\frac{-2\log_e S}{S}}$$

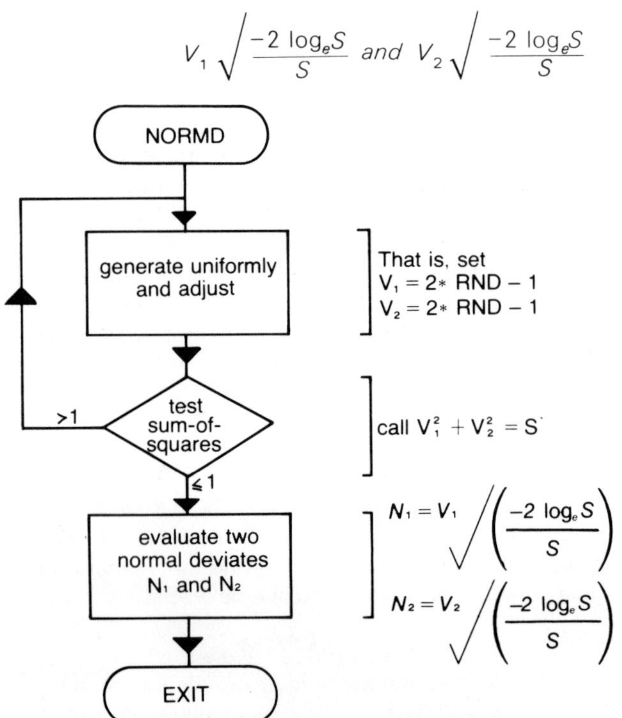

Fig. 44 An algorithm returning values from a Normal distribution

are suitably distributed; the theory is beyond the scope of the text.

Negative exponential random numbers

In simulating, for example, a queue at a supermarket we should expect a customer to arrive, on average every *m* seconds. The time between the arrival of one customer and the next—the waiting time, is described by random numbers chosen from the negative exponential distribution. We find considerably more short time periods than long ones in the random times given out by the following method.

We use the logarithm function LOG described in the previous chapter. Where customers do arrive on average every *m* seconds, a suitable generator of random times is given by

```
100 LET X = −M∗LOG(RND)
```

Remember that the values produced by RND are all in the range 0 to 1, so that their logarithms will be negative. Thus the statement above will produce positive values only, as is appropriate! There is a slight chance that this algorithm might produce an overflow warning message if the value given by RND is very small, and hence LOG(RND) is large and negative.

Poisson-distributed random integers

The Poisson distribution is closely connected to the exponential distribution of the previous section. A Poisson-distributed random number would give the number of 'events' (people joining the queue for example) in a given unit of time, where the time between events is given by the exponential distribution. Alternatively, it stands in its own right as descriptive of situations such as telephone exchanges where the probability of an 'event' (a particular phone being called) is low, but there are a large number of participants (a lot of telephones).

The algorithm given in Fig. 45 produces integers from a Poisson distribution. The value of *m* is the same as the value of *m* in the corresponding exponential distribution. It is worth observing that the integer returned from this algorithm is the number of times the algorithm loops until the condition is satisfied. For large values of *m*, this will result

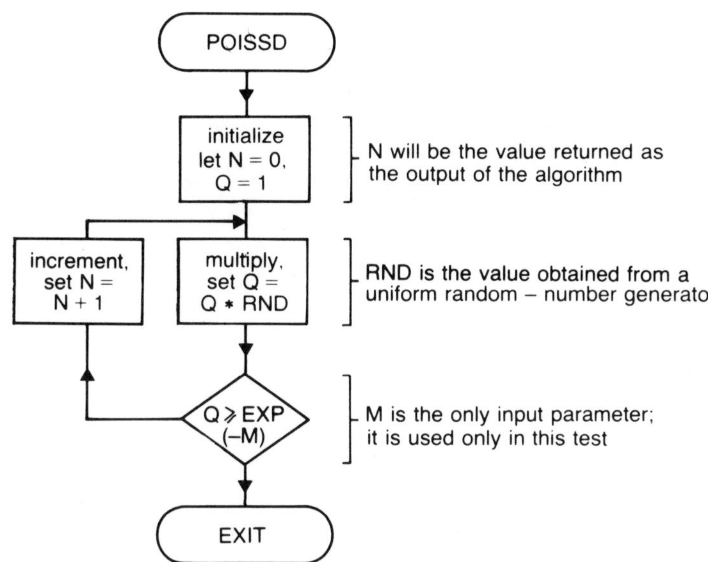

Fig. 45 An algorithm returning values from a Poisson distribution

in an unacceptable increase in processing time, so this algorithm is best restricted in use to small values of *m*.

Summary

In this chapter we have explored the nature of random sequences of numbers. We have observed that they are of use in simulating a wide variety of events, but that the process of verifying their 'randomness' is a process of applying a battery of tests, each one of which must be satisfied.

Computers use pseudo-random number generators, producing a determinate sequence of numbers that appear to satisfy these tests. However, if restarted from the same point, the same numbers will result.

Two principal methods are used in practice to generate random numbers: the middle-square method and the linear congruential method. The latter has the advantage that it is not so critically dependent on its starting value as the middle-square method, and indeed it has four constants whose values may be chosen to optimize the results.

Finally, methods of manipulating the results returned by these methods, in order to obtain useful sequences, are examined. The detailed justification is not given, being more properly found in a book on statistics. Some suggestions are given in the references at the end of the book.

EXERCISES

Chapter material

1. *If* r_1 *and* r_2 *are two uniformly-distributed integers in the range 0–N, can we get a 'more random' integer by forming the product* $r_1.r_2$*?*

Public examination questions

2. *A function RANDOM has an effect which is dependent on its argument, as follows:*

RANDOM(X)	*X > 0: the pseudo-random number generator is set to an initial state determined by the value of X; 0 is returned;*
RANDOM(X)	*X < 0: the pseudo-random number generator is set to an initial state determined by the time of day; 0 is returned;*
RANDOM(X)	*X = 0: the next value from the pseudo-random number generator is returned; this value is between 0 (inclusive) and 1 (exclusive), and a large enough sample of such values will be uniformly distributed across the range.*

   a. *Explain why it is useful to have the control over a pseudo-random number generator which is provided by the calls with a non-zero argument as described above.*
   b. *Describe how the function RANDOM may be used to provide a sample of pseudo-random numbers uniformly distributed across*

the range of values 1, 2, 3, 4, 5, 6. (You may assume the existence of other simple arithmetic functions.)

c. *An estimate is required of how many times on average a die must be thrown before each of the scores 1 to 6 is obtained at least once, and a simulation of 1000 trials is to be carried out. Either describe a suitable algorithm, by means of a flowchart or otherwise, or write a program for the task in a language of your choice, annotated to aid the reader.*

(Cambridge Specimen Paper) (8%, 3 hours)

Projects

3. *Investigate the middle-square method for all starting values from 0000 to 9999 to see which eventually cycle on 0000.*
4. *Investigate the middle-square method for starting values from 00 to 99 to see which cycle in less than the full period.*
5. *Apply the tests of Section 6.1 to the random-number generators supplied with your high-level software. Are you satisfied? If not, do Exercise 6.*
6. *Write a better random-number generator. You will need to work in assembly language.*

7 Errors

In English, the word 'error' means mistake, accident, failing . . . and so on. This use carries with it the idea that errors could be avoided with sufficient care.

This is not the case in the computing (more correctly the mathematical) use of the word. Errors in computing arise almost entirely because we are doing arithmetic with finite length words. They are unavoidable, but this chapter is about how to know when to expect them, and how best to live with their effect. However careful we are, we cannot avoid introducing error into our numerical computing.

Just as no one intentionally commits errors, no one intentionally exagerates them; but in computing, the fundamental operations that we perform on numbers have exactly this undesirable effect. Indeed, there are two kinds of computer programmer: the kind that believes that 2 + 2 = 3.999999, because the computer tells him so, and the kind that believes that we shall never know what 2 + 2 equals, because any answer is almost bound to be in error! This chapter is directed at these programmers, in the hope that they can learn to live together.

7.1 What is an error?

If, in denary, we attempt to write down the fraction 1/3 as a decimal, the best we can do is

0.33333333333333 . . .

(The alternative notation $0.\dot{3}$ doesn't help us much in what follows, so we shall ignore it.)

This is not the true value of 1/3, which could only be written out correctly by use of an infinite number of 3's. However, in writing down this decimal, and similarly in storing it in a finite length computer word, we must stop somewhere. Wherever we stop, we shall have introduced an *error*. This error is the difference between the true value, and the value we have stored. Thus

0.3	introduces the error	0.033333333 . . .
0.33	introduces the error	0.003333333 . . .
0.333	introduces the error	0.000333333 . . .
. . .	. . . . . .	. . .
0.333333	introduces the error	0.000000333 . . .

There is a problem here, because we cannot say exactly how much the error is (once again, we would need an infinite number of 3's). Sometimes, we cannot say exactly what the error is because we don't know what the true answer should be. For example, if we are told that the result of a lengthy calculation, given to five decimal places is 0.12345, how can we know how much in error the answer is—it could be exactly right, it could be an answer that should be 0.12345333333333 . . ., it could be any one of a limitless range of possibilities.

But there is something useful that we can say in each of these situations. We can set a limit on the size of the error. The error in representing 0.333333333 . . . as 0.33333, that is, 0.000003333 . . . is certainly less than 0.000004, and *that* is a definite quantity that we can write down. Similarly, whatever the true result represented by 0.12345 is, it must be smaller than 0.123455 (or else the 5-place result would be different) and similarly must be greater than 0.123445. The result cannot be too small by a greater amount than 0.000005 nor too large by more than 0.000005. The error is certainly less than 0.000005.

So we do not work with errors as such: we work with the *error bound*, which is the maximum size of the error.

By convention, the error bound is always positive, so if we say that a result is 0.33333 with an error bound of 0.000004, we are actually saying that the true answer lies somewhere in the range from 0.33333 − 0.000004 up to 0.33333 +0.000004, that is between 0.333326 and 0.333334. Sometimes we write this as 0.33333±0.000004. An excellent rule for programmers is always to present their numerical results with some indication of the likely error bound.

7.2 How do errors occur?

The commonest source of errors in computing is the storage of a number in a fixed-length word. Our denary examples in Section 7.1 have been just two of the possible examples. It is worth looking at some of the situations where errors can arise, in more detail.

Integer storage

Where integers are being stored as fixed-point representations, as in Section 1.1, the error bound is zero. That is, integers can be stored exactly. This of course is no longer true when overflow occurs. With 8 bits, using any of the signed conventions (see Section 1.1), numbers larger than 128 will be in error, and the error will increase as the size of the number to be stored increases. Indeed, most systems will not allow this to happen without a warning message.

Fixed-point fractional storage

Although a few fractions may be represented exactly, as fixed-point quantities (see Section 1.2), the majority cannot. We can say quite easily though what an error bound is, once we know the word-length. With the (binary) point to the left of the second bit, and using 8 bits, the least significant bit represents 2^{-7}; therefore the error bound is 2^{-7}. If the error was greater than this, the last digit of the representation would be affected. We might actually be able to be more precise, and give a smaller error bound in certain fortunate cases, but whatever the number being *stored*, the error bound cannot be worse than 2^{-7}. It may well be considerably worse after a series of calculations, as we shall see in Section 7.3. This is true, regardless of whether the computer rounds or truncates.

Floating-point storage

Some quantities may be represented accurately in floating-point form. Again, the error bound in other situations can easily be calculated. It depends on the word-length, specifically on the amount of storage allocated for the mantissa, and also on the size of the exponent. The error bound where 8 bits are used in fixed-point fractional form for the mantissa (cf., Section 1.3) is

$$2^{-7} \times 2^{\text{EXPONENT}}$$

It is sometimes more convenient to deal with *relative error bounds* rather than error bounds. The *relative error* is simply the (absolute) error, as before, divided by the actual stored result. It is easy to give a relative error bound for a floating-point representation. In our example, the relative error is

$$\frac{2^{-7} \times 2^{\text{EXPONENT}}}{\text{F} \times 2^{\text{EXPONENT}}}$$

where F is the (normalized) mantissa; as this lies between $\frac{1}{2}$ and 1, in the worst possible case, the relative error will be $2^{-7}/\frac{1}{2} = 2^{-6}$. Thus, the relative error bound for all floating-point representations, using 8-bit mantissae, regardless of the size of exponent, is 2^{-6}. We shall see later that certain of the rules for keeping track of errors are more easily memorized using relative bounds rather than absolute error bounds. Once again, it is customary to ignore the sign of relative errors.

7.3 How do errors grow?

Given two computer representations of numbers, both potentially in error, it is possible that their addition, or subtraction, or multiplication, or whatever, causes the errors to cancel out, so that we get an exact answer, even though the operands were in error when they were stored. It is doubtful, however, if we would know when this occurred, and indeed, we might be so unlucky that the errors reinforced one another. Because of this possibility, it is clear that the error bound after an operation must be larger than either error bound before. The whole point of error bounds is preparation for the worst possible case; we can then only be pleasantly surprised!

The error bound of a sum

Let us suppose that we have two numbers X_{true} and Y_{true}. When they are stored as X_{store} and Y_{store}, there will be two errors: e_X and e_Y. Suppose, in the worst possible case, that X_{true} and Y_{true} are both greater than their stored representations, that is

$$e_X = X_{true} - X_{store} \text{ and } e_Y = Y_{true} - Y_{store}$$

Then clearly

$$(X_{true}+Y_{true}) = (X_{store}+Y_{store}+e_X+e_Y)$$

Now $(X_{true}+Y_{true})$ is the correct value of the answer, so the error after addition has grown to

$$e_{X+Y} = e_X + e_Y$$

This is the first rule of the 'propagation of errors':

The error bound of a sum is the sum of the individual error bounds

Again, remember that it could be that one stored value is greater than the true value, and the other less: in that case, the result will be closer to the true result than our pessimistic view. However, we are looking at worst possible cases.

The error bound of a difference

The reader can construct a worst possible case for himself. We do not subtract the errors, we add them, even though the operation is subtraction and not addition. The rule is:

The error bound of a difference is the sum of the individual error bounds

The error bound of a product and a quotient

We shall not spend time deriving these; the mathematically-inclined may care to derive them as indicated in the exercises. The rules are as follows: (notice that the errors in the rules are *relative* errors)

The relative error bound of a product is the sum of the individual relative error bounds

The relative error bound of a quotient is the difference of the individual relative error bounds

For the mathematically inclined

There is a close and fairly obvious link between the calculus and the propagation of errors as described above. By treating x^n as the n-fold product of x, and converting between relative and absolute errors, we can show that the absolute error in x^n is given by

$$e_{x^n} = n.x^{n-1}.e_x$$

which is reminiscent of the formula for the derivative of x^n

The error in polynomials

By use of the rules given above, it is possible to calculate the error bound in evaluating any polynomial, by treating it as a sequence of exponentiations, multiplications, additions or subtractions.

7.4 Some common error-prone situations

The preceding sections have been largely theoretical. The question arises—when can we expect errors, and what situations should we avoid, in order to minimize error?

The following is a check list of some of the sources of error in computer algorithms.

Insufficient precision

We have already observed, in our consideration of the sine algorithm, in Section 5.4, that large numbers in floating-point format lack enough digits of precision to enable sensible calculations to be performed on them. The same situation can arise when we are dividing by a very small quantity. We shall have more to say about this in the section on non-commutative arithmetic, but as a general rule, if it is possible to alter the order of calculation in order to render division as an operation between operands of approximately the same magnitude, this will be to the good.

Loop variables

It is tempting to initiate a BASIC FOR . . . NEXT loop with the statement

```
100 FOR I = 1 TO 2 STEP 0.1
```

Because 0.1 is a denary quantity that has no exact decimal equivalent to any degree of accuracy, we may find that this loop is not executed as many times as we expect. This is due to the fact that in actually performing the loop, the step size is only approximated, and subsequent additions of this step will exagerate the error. It may be more profitable to program the loop as

```
100 FOR J =10 TO 20 STEP 1
110 LET I =J/10
... .......
```

which will keep the error in I to a minimum, and ensure that the loop is executed the expected number of times; it is well worth using tricks to ensure that any loop is controlled by an integer.

Non-commutative arithmetic

We may well find that the result of the two statements

```
    100 LET X = (A*B)/(C*D)
and 100 LET X = (A/C)*(B/D)
```

is different. If the operands A,B,C,D are large, the product A*B may well be so large that precision is lost, and similarly with C*D; whenever possible, rearrange a calculation to keep floating-point quantities around the same order of magnitude.

Subtraction of nearly equal floating-point quantities

Normalization (Section 1.3) is a double-edged weapon. In Section 4.6 we give an example to show that the subtraction of nearly equal floating-point quantities can result in a dramatic loss of precision. If a calculation can be rearranged to avoid such a subtraction, this should be done.

Non-equality of floating-point quantities

Because of the possibility of accumulated errors due to arithmetic operations, described earlier in the chapter, it is rare that two quantities held in floating-point storage will be exactly equal, even if they should be 'mathematically' equal. It is very unwise to hinge a program on a statement such as

```
100 IF A = B THEN 500
```

where either, or both, of A and B are arrived at as the result of a computation. A better approach is to select some 'tolerance' and to test for 'equality within a tolerance' as below:

```
100 LET E =0.000001
110 IF ABS(A-B) < E THEN 500
```

Factorials in power series

In a power series, such as those discussed in Chapter 5, the factorial function $x!$ often appears. Thus:

$$\sin(x) = x - \frac{x^3}{3!} + \frac{x^5}{5!} - \dots$$

Although nested multiplication is one possibility for evaluating this series, another way, minimizing the effect of division by the rapidly growing factorial function (see Section 5.10), is to generate each new term from the previous by a statement such as

```
100 LET T1 =T*X*X/(N*(N-1))
```

which could perhaps be rearranged to preserve precision.

Reducing error

For completeness, we should perhaps mention the possibility of reducing error by the use of *extended precision*. This is usually available through software, rather than hardware. Thus, if an 8-bit word does not carry enough precision for the programmer to represent fixed-point fractions to his satisfaction, a further 8-bit word may be utilized, carrying a further 8-bits' worth of accuracy. All arithmetic operations now become more complicated, and must be programmed explicitly. Cross products during multiplication, for instance, or carries from one word to another during addition must be accounted for, and the usual result is a considerable increase in processing time.

For operations on very large numbers, it is possible to hold individual (denary) digits as entries in an array. Again, the arithmetic operations must be explicitly programmed, and a severe penalty in terms of processing time is exacted. It is however, only in this manner that large-scale computations, such as proving that $2^{11213}-1$ is a prime number, can be carried out.

Summary

In this chapter, we have shown that error, in the computing sense, is an unavoidable consequence of finite length words. It is accordingly appropriate to rearrange numerical calculations wherever possible to avoid the computations listed in Section 7.4, and also to attempt to give some sort of error bound on an answer.

Much of the detailed mathematical analysis of error is beyond this book, and may be found in texts on numerical analysis, such as those mentioned in the bibliography.

EXERCISES

Chapter material

1. *How many fixed-point fractions can be exactly represented in 8 bits?*
2. *How many floating-point quantities can be exactly represented in 2, 8-bit words?*
3. *Derive the results given in Section 7.3 for the error bound of a product and a quotient, for example, for multiplication take*

$$e_X = X_{true} - X_{store} \qquad e_Y = Y_{true} - Y_{store}$$

Rearrange these to give expressions for X *and* Y *and multiply out. Inspect the answer. You will need to ignore the product* $e_X.e_Y$ *(because it is small enough to be neglected).*

4. *Find a factor of* 2^{11213}
5. *A computer is programmed to compute the value of the expression* $X^2 - Y^2$
   a. *by direct evaluation of* X*X −Y*Y
   b. *by evaluation of the mathematically equivalent expression* (X −Y)*(X +Y).

   When X *has the value 10 000 and* Y *the value 9999.9, the first method gives the result 2000.00 and the second 1999.99. Comment on this discrepancy.*
   (London, 1976) *(≃ 4%, 3 hours)*

Projects

6. *Prove that* $2^{11213} - 1$ *is prime (check that you have sufficient computer time!)*

Answers

Chapter 1

		a.	b.	c.
1.	(i)	01010011	01010011	01010011
	(ii)	11010011	10101101	10101100
	(iii)	10000001	11111111	11111110
	(iv)	00011111	00011111	00011111

		a.	b.	c.
2.	(i)	−127	−1	−0
	(ii)	+127	+127	+127
	(iii)	−0	−128	−127
	(iv)	−42	−86	−85
	(v)	−77	−51	−50

3. a. +4095, −4095 b. +4095, −4096 c. +4095, −4095

4. 10 bits, since $2^9 < 1000$ and $2^{10} > 1000$

		a.	b.	c.
5.	(i)	−0.9921875	−0.0078125	−0
	(ii)	+0.9921875	+0.9921875	+0.9921875
	(iii)	−0	−1	−0.9921875
	(iv)	−0.328125	−0.671875	−0.6640625
	(v)	−0.6015625	−0.3984375	−0.390625

6. (i) 0 1100000
(ii) 0 1010000
(iii) 0 0000100
(iv) 1 0110000
(v) 1 1000000

7. (i) 0 0001101 (= 0.1015625)
(ii) 0 0100110 (= 0.296875)
(iii) 0 0101011 (= 0.3359375)
(iv) 1 1110011 (= −0.1015625)
(v) 1 1100110 (= −0.203125)

8. a. 0 11111111111 1111 approx. 1×2^7
b. 0 10000000000 0000 $0.5 \times 2^{-8} = 2^{-9}$
c. 1 00000000000 1111 -1×2^7
d. 1 01111111111 0000 approx $-0.5 \times 2^{-8} = 2^{-9}$

9. a. approximately 128
b. approximately 1.9531×10^{-3}
c. −128 (exactly)
d. approximately -1.9531×10^{-3}

10. a. 0.8515625×2^{42} approx 3.7452×10^{12}
b. $-0.8984375 \times 2^{127}$ approx -1.5286×10^{38}
c. -0.8828125×2^{29} approx -4.7396×10^{8}

Chapter 3

		a.	b.	c.
1.	(i)	11110000	00010000	00010000
	(ii)	11111000	11111000	11111011
	(iii)	00000000	00000000	11111111
	(iv)	11111111	00000000	11111111
	(v)	00001111	00001111	00001111

2. (i) 11111110
(ii) 11110001
(iii) Overflow (01111100)
(iv) Overflow (10000000)
(v) 00000000

CHAPTER 6

1. No. There is only a 50% chance that r_1 is even and similarly for r_2. But there will be a 75% chance that $r_1.r_2$ is even. The product is less random than either of the factors.

CHAPTER 7

1. 256, for the usual reason $2^8 = 256$. There will be 128 negative ones and 128 positive (including zero).
2. The answer now is not simply 2^{16}, because of repetitions. If the quantities are normalized, one of the bits is effectively fixed so that we

have 2^{15} *different normalized repetitions. None of these is zero, which adds an extra one:* $2^{15} + 1 = 32769$

3. $X_{true} = X_{store} + e_X$
$Y_{true} = Y_{store} + e_Y$
$$X_{true} \cdot Y_{true} = (X_{store} + e_X).(Y_{store} + e_Y)$$
$$= X_{store} \cdot Y_{store} + e_X \cdot Y_{store} + e_Y \cdot X_{store} + e_X \cdot e_Y$$
$$\simeq X_{store} \cdot Y_{store} + e_X \cdot Y_{store} + e_Y \cdot X_{store}$$
$$e_{XY} \simeq e_X \cdot Y_{store} + e_Y \cdot X_{store}$$
$$r_{XY} = \frac{e_{XY}}{X_{store} \cdot Y_{store}} \simeq \frac{e_X}{X_{store}} + \frac{e_X}{Y_{store}}$$
$$= r_X + r_Y$$

and similarly for division.

4. *2!*

Further Reading

This short bibliography consists of books that explore the topics of this book in somewhat greater depth. There will, of course, be many alternatives to the books on this list. However, it is hard to see what could possibly replace:

The Art of Computer Programming D. E. Knuth (Addison-Wesley)
Vol I *Fundamental Algorithms* (2nd edition 1973)
Vol II *Seminumerical Algorithms* (1969)

For much of the material in Chapter 2, I am indebted to:

Computer Technology for Technicians and Technician Engineers R. V. Watkin (Longman, 1976)
The Digital Computer Open University Course TM221 (Open University Press).

For the mathematical background to Chapters 5 & 7:

Numerical methods and FORTRAN Programming D. D. McCracken & W. S. Dorn (Wiley, 1964)
Introductory Computer Methods and Numerical Analysis R. H. Pennington (Macmillan, 1966)

For the statistical background to Chapter 6:

Introduction to Mathematical Statistics P. G. Hoel (Wiley, 1962)

Index